BIOLOGY for TODAY

Book 2

Ernest G. Neal

MBE, M.Sc., Ph.D., F.I.Biol.

formerly Second Master and Head of the Science Department
Taunton School

and

Keith R. C. Neal

M.A., F.I.Biol.

Head of the Biology Department
Manchester Grammar School

Line illustrations by Barry Jones and Marion Mills

Blandford Press   Poole   Dorset

First published in the U.K. 1974 by
Blandford Press Ltd,
Link House, West Street,
Poole, Dorset BH15 1LL

Reprinted (with corrections) 1976
Reprinted 1978
Reprinted 1980

ISBN 0 7137 0661 9 (School Edition)
ISBN 0 7137 0675 9 (Library Edition)

Filmset by Keyspools Ltd, Golborne, Lancs
Printed in Great Britain by
Butler and Tanner Ltd, Frome, Somerset

Authors' Preface

In recent years the publication of enquiry-based courses in Biology has greatly stimulated the teaching of the subject, and these have already led to a welcome change of emphasis in the traditional courses. We recognise the great importance of the enquiry approach and believe that many topics are admirably suited to this method; however, we feel that other topics are better explained in a more traditional manner. Therefore we have not hesitated to incorporate both these approaches in the course. It is also our belief that a text-book at this level should provide a clear framework of easily accessible factual information and that the text, illustrations and practical work should be closely integrated in order to lead the student towards a clear understanding of biological principles.

The course is designed to meet the requirements of the syllabuses of the various Examining Boards for the General Certificate of Education at Ordinary Level in Biology.

We are convinced that any O-level course in Biology should be educational in the widest sense of the word and should provide future citizens with a better understanding of themselves and their environment. For this reason considerable emphasis has been given to Man and to the application of biological principles to human affairs. We have also given special prominence to subjects such as behaviour, ecology and conservation because of their great relevance to life in the 1980's.

The arrangement of topics in a Biology course is always a matter of individual preference. This course is divided roughly into three sections. In Book 1 the first 11 chapters introduce the student to basic biological principles through the study of a wide variety of animals and plants. The remaining chapters are primarily functional in approach and deal in greater detail with Man and flowering plants. This is continued in the first 9 chapters of Book 2. The final section is concerned with wider topics including Man and his environment, genetics and evolution, and draws together many of the principles from previous chapters.

Throughout the course we have kept in mind the changing interests and powers of comprehension of the student which are associated with increasing age and experience, but the order can easily be adjusted to suit the individual preferences of the teacher.

We have tried to include adequate instructions in the text for the practical work to be carried out successfully, but some additional information, together with suggestions for relevant teaching aids, has been included in the Appendix.

Nobody can write a text-book without being greatly indebted to many people. We are very conscious of the great debt we owe to the authors of the many works we have consulted and would specially mention those concerned with the Nuffield Science Teaching Project and the Scottish Course, Biology by Inquiry. Some of the experiments and ideas in this book have been based on these and other courses. But an author's gratitude can only partly be covered by formal acknowledgments because equally important are the many helpful discussions and influences which can less easily be attributed to an individual.

Particular thanks must go to Mr John Haller of Philip Harris Biological Ltd., who has given most practical help in the provision of photographs, many of them taken by himself, and for his enthusiastic cooperation over the whole project. The line illustrations have been specially drawn by two artists, Barry Jones and Marion Mills. We are very grateful to them for their close cooperation over each diagram and for producing such outstanding work. A detailed list of acknowledgments relating to the illustrations will be found on p. 191. It is regretted if any credit has been unwittingly omitted.

We are most grateful to the following who have read sections of the manuscript and given us the benefit of their advice, criticism and encouragement: Dr David Bygott, Mr Howard Green, Miss Muriel Hosking, Mr Martin Jacoby, Mr Andrew Neal, Dr Philip Penny, Mr Gordon Perry, the late Mr Colin Russell, Mr Geoffrey Stephens and Dr David Watson; also Mrs Hazel Watson and Mrs Elizabeth Wells for typing the manuscript. Finally, we warmly thank our wives for reading the manuscript critically, for their many valuable suggestions and for their patience and encouragement.

E.G.N.   K.R.C.N.

Contents

1

Transporting materials within the animal

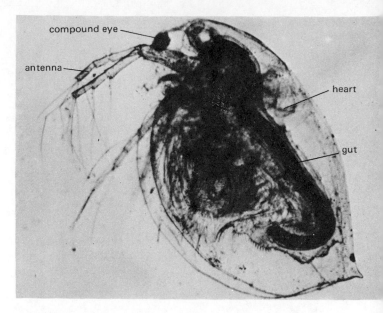

Fig. 1:1 A water flea (*Daphnia*) ×18.

Every cell in the body has to have a continual supply of essential nutrients and oxygen in order to carry out its metabolic activities, and a means of removing the excretory substances which are formed within it. A brain cell, or a cell in the big toe, is a long way from the food which is digested in the gut and from the oxygen which enters the body through the lungs. Hence a transport system is essential for connecting up all the cells, wherever they may be, with the sources of supply of these vital substances. In the larger animals the **blood system** is the main transport system of the body.

BLOOD SYSTEMS

The simplest way to see how a blood system works is to look at an animal which is fairly transparent. One of the best for this purpose is the water flea, *Daphnia*, but small specimens of the water louse, *Asellus*, or the water shrimp, *Gammarus*, are good substitutes.

Place a small *Daphnia* in a few drops of water on a cavity slide, lower a cover slip carefully and examine it under the medium power of a microscope. Locate the position of the heart (Fig. 1:1). Observe the movement of the heart carefully. Can you see any tubes coming out of the heart? Now turn to the high power and focus on the area near the heart and look out for any particles which are moving about.

Do these particles flow along smoothly or in jerks? Do they move in a particular direction or in a haphazard manner?

These minute particles are blood corpuscles and their movement is due to the pumping action of the heart. In Arthropods such as *Daphnia* the blood is not confined in tubes, but is pumped through a series of spaces which surrounds the organs. When blood returns to the heart it passes through openings in the heart wall which are controlled by valves. This type of circulatory system is called an **open system**. In the majority of animals, including ourselves, the system is said to be **closed** as the blood travels within tubes—the **arteries** and **veins.**

It is not easy to find out how the blood is flowing in our own blood system as we are not transparent like *Daphnia*, so it is not surprising that up to the 16th century people believed that blood flowed backwards and forwards like the tides.

William Harvey (1578–1657) first demonstrated that blood actually circulated through arteries and veins in a particular direction. By many dissections and ingenious experiments he showed that the blood in arteries flowed away from the heart and that in veins towards it. He demonstrated the flow of blood in veins in this way. First he tied a band tightly round a man's upper arm which caused the veins in

1

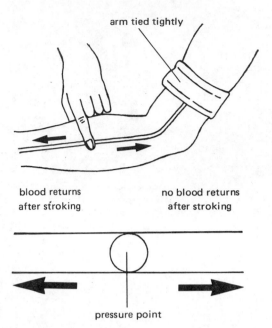

blood returns
after stroking

no blood returns
after stroking

pressure point

Fig. 1:2 Diagram showing the principle of Harvey's experiment. Pressure is exerted on the vein with one finger while with another finger the vein is stroked firmly up the arm and then down the arm.

his arm to become prominent. He then placed one finger firmly on the middle of one vein and kept it there throughout the experiment. Then, with another finger, he squeezed out the blood from the part of the vein *above* this pressure point by stroking it firmly *towards* the upper arm. On releasing the pressure the blood did not flow back and fill the vein. When he repeated the action on the part of the vein *below* the pressure point, squeezing *down* the arm, on releasing the pressure this time the blood *did* flow back into the vein (Fig. 1:2). So he concluded that the blood in veins flowed towards the heart.

Harvey was unable to demonstrate any connection between the smallest arteries and the smallest veins, but a few years later an Italian, **Marcello Malpighi** (1628–94), using a better quality lens, was able to see fine tubes connecting the ends of minute arteries with veins in the lung of a frog. These microscopic connecting tubes are called **capillaries.**

BLOOD CIRCULATION IN MAN

This consists of a heart which pumps the blood round, arteries which take blood away from the heart, veins which bring it back to the heart

and capillaries which join the smallest branches of the arteries, the **arterioles,** to the finest branches of the veins, the **venules.**

How is it that every living cell in our body obtains the vital supplies that it needs? If the *same* blood went to all the organs, the substances it carried would soon be used up and the last organs to be reached would have none. For example, oxygen enters the blood at one place only—the lungs—but every organ needs oxygen, so in some way fresh blood has to reach each organ. This is brought about because each organ has a special circuit of its own consisting of an artery and vein and their branches, and the capillaries which join the arterioles with the venules. The only exception to this is the circulation to the lungs because *all* the blood goes through this circuit and becomes oxygenated, and on its return to the heart it is sent to all the other circuits and finally back to the heart again. You will see from Fig. 1:3 that because of this special circuit to the lungs the blood passes through the heart twice for each time it flows through any other part; this is known as a **double circulation**. Note that on one circuit the blood passes through the right side of the heart, but on the other circuit it passes through the left side.

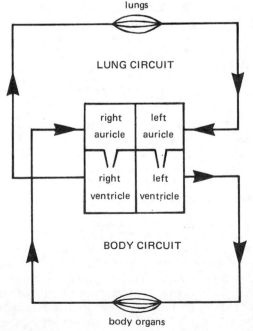

lungs

LUNG CIRCUIT

| right auricle | left auricle |
| right ventricle | left ventricle |

BODY CIRCUIT

body organs

Fig. 1:3 Schematic diagram showing the principle of the double circulation.

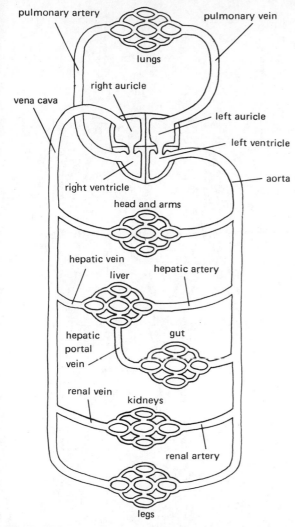

pulmonary artery

pulmonary vein

lungs

right auricle

vena cava

left auricle

left ventricle

right ventricle

aorta

head and arms

hepatic vein

hepatic artery

liver

hepatic portal vein

gut

renal vein

kidneys

renal artery

legs

Fig. 1:4 The main blood circuits of the body (schematic).

Compare Fig. 1:4 with Fig. 1:5. The former shows the main circuits in highly diagrammatic form, the latter the natural positions of the main arteries and veins making up these circuits. In most circuits the artery and vein run alongside each other until they break up into finer branches. The main artery of the body, the **aorta**, runs in the mid-line alongside the main vein, the **vena cava**. The circuit concerned with the supply of blood to the gut is rather different from the others as the blood on leaving the gut does not return to the vena cava direct but passes first to the liver via the **portal vein**. This is the only vein in a mammal which has capillaries at both ends.

Why is it functionally desirable that blood from the gut should go direct to the liver?

Examine a dissection of a rat or rabbit and identify as many as possible of the main arteries and veins.

We shall now study the component parts of the circulatory system in more detail.

The heart

The driving force of the circulation is, of course, the heart. Functionally the heart can be thought of as an organ composed of left and right sides, each side acting as a separate pump for the two circulations. Oxygenated blood from the lungs enters the left **auricle**, passes through a valve into the left **ventricle** and is expelled via the aorta to the head and body. On the other side of the heart, deoxygenated blood enters the right auricle, passes through a valve into the right ventricle and is then pumped through the pulmonary artery to the lungs.

Examine the heart of a sheep, pig or cow. All are basically similar to the human heart. Study Fig. 1:6 and try to identify the various parts including the blood vessels.

Also examine a heart which has been dissected to display the internal structures. Note that the valves between the auricles and ventricles are connected to pillar-like muscles in the walls of the ventricles by tough, non-elastic cords. These hold the valve flaps in position so preventing blood from flowing back into the auricles when the ventricles contract. If the aorta and pulmonary arteries have not been completely removed, you should be able to see inside them the semi-lunar valves which prevent blood from flowing back into the ventricles after the beat. These valves work on the same principle as the pocket valves in veins (Fig. 1:8d). Compare the thickness of the walls in all four chambers of the heart. Can you think of any reasons for the differences?

The heart beat

This is a double action. First the two auricles contract, forcing blood into the ventricles, then the ventricles contract immediately afterwards, pumping the blood into the arteries. Between

3

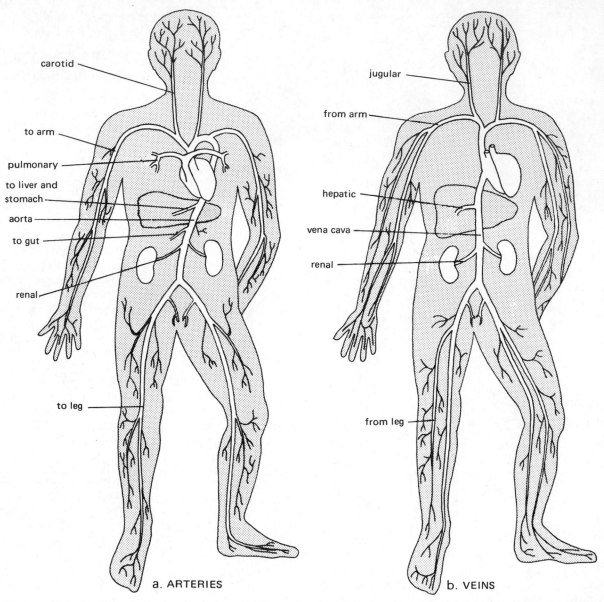

carotid

to arm

pulmonary

to liver and
stomach

aorta

to gut

renal

to leg

a. ARTERIES

jugular

from arm

hepatic

vena cava

renal

from leg

b. VEINS

Fig. 1:5 The blood circulatory system in man: a) the main arteries b) the main veins (the portal vein is not shown).

these two contractions, the valves between the auricles and ventricles close so that the blood is directed only into the aorta and pulmonary artery when the ventricle walls contract. At the same time as the ventricles contract the auricles relax and fill up again.

The rate at which the heart beats in an adult averages 72 beats per minute. Find out your own rate by taking your pulse, or ask your partner to take it for you.

The pulse in the wrist is the easiest one to take. Rest your left arm on the bench with the hand facing upwards, then place the tips of the first and second fingers lightly on the thumb side of the wrist. Count the number of beats in a 30 second period. Repeat two or three times and obtain an average figure. Calculate the number of beats per minute and compare your rate with the rest of the class. There will probably be considerable variation. How does the average

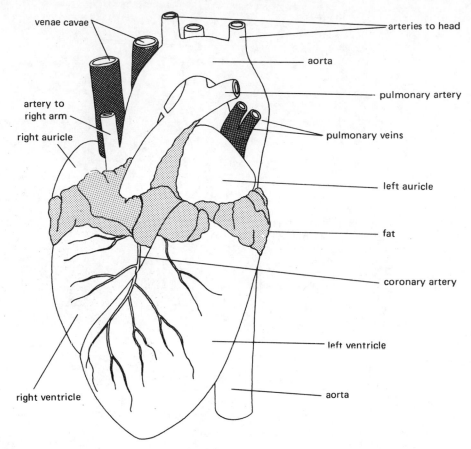

venae cavae

arteries to head

aorta

pulmonary artery

artery to
right arm

pulmonary veins

right auricle

left auricle

fat

coronary artery

left ventricle

right ventricle

aorta

Fig. 1:6 The external features of a sheep's heart and its associated vessels, ventral view

rate of the class compare with the average adult rate?

You will know that your heart beats faster during and after exercise. You could find out what variation there is between members of your class in the way the heart reacts to exercise.

In order to make the comparisons more exact, it is best for the class to take the same amount of vigorous exercise. As soon as the exercise is finished, take your pulse rate continuously over half-minute periods until it has returned to the original figure. You should plot a graph of your own pulse rate against time. Now compare the graphs of the whole class and find out:
1. What variation in recovery time is there between individuals? Do those with the lowest initial rates recover faster or not?

2. In how many cases does the rate fall below the original? Can you suggest any possible reason for this?
3. Compare the results of those who train regularly for games, athletics or swimming with those who do not over-exert themselves. Can you draw any conclusions?

The mechanism by which the heart beat is regulated according to the overall needs of the body is controlled through nerves and chemical hormones (p. 109).

When we are active the body responds by pumping more blood to the muscles. This, together with an increased breathing rate, ensures that more oxygen and food reaches the muscles and that waste products are removed quickly.

5

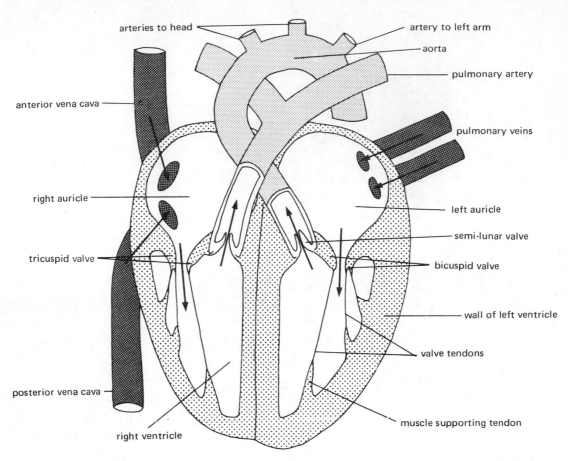

Labels on figure:
arteries to head
artery to left arm
aorta
pulmonary artery
anterior vena cava
pulmonary veins
right auricle
left auricle
semi-lunar valve
tricuspid valve
bicuspid valve
wall of left ventricle
valve tendons
posterior vena cava
muscle supporting tendon
right ventricle

Fig. 1:7 Simplified diagram of a longitudinal section through the human heart. The arrows show the direction of blood flow.

Arteries

The system of arteries is rather like a tree which divides into smaller and smaller branches. When the heart beats it forces the blood into this system, but there is considerable resistance to its passage and great arteries near the heart take most of the pressure. Their walls are consequently very strong but also elastic (Fig. 1:8a) so that they can dilate and take up the increased volume of the blood when the ventricles force more blood into them. Immediately afterwards, when the ventricles relax, the elastic walls of the great arteries return to their original diameter and force the blood along. This wave of expansion and contraction becomes less the further away the arteries are from the heart; this is because the walls of the more distant arteries are more muscular and less elastic. However, this wave can still be felt in the wrist as a pulse which, if counted, indicates the heart rate. You could, of course, take your pulse in many parts of the body if you could feel the appropriate artery easily. Try this for one of the leg arteries:

Sit with one leg firmly on the floor and the other resting on it so that the back of one knee rests on the knee of the other. After a time you will see and feel the leg which is on top give a series of small movements with each heart beat. If you do it for long you will reduce the blood flow to the leg and so develop 'pins and needles'.

The importance of the muscle layer in the *smaller* arteries and arterioles is that blood flow can be controlled according to their degree of contraction, thus some blood circuits can be given more or less blood depending on the demands of the body. You will see later how this affects the circulation to the skin (p. 78) and the gut (p. 110).

Veins

By the time the blood from the heart reaches the veins, the pressure is extremely low, a fact which accounts for their relatively thin walls. Veins depend largely upon movements in the surrounding muscles for squeezing the blood back to the heart. The larger veins contain valves to ensure that blood cannot travel in the wrong direction (Fig. 1:8d). Valves are especially important in veins where the blood flows against gravity, as in the arms and legs.

You can find out the position of the valves in your veins by examining your arm:

Swing your arm round several times to fill the veins with blood, hold the arm vertically downwards and gently press your finger along a prominent vein—stroking it in the reverse direction to the blood flow, i.e. towards the hand. Can you see the swellings where you have pushed blood against the valves?

Capillaries

These microscopic vessels which join arterioles to venules are the *only* parts of the circulatory system where substances can enter or leave the blood (Fig. 1:8c). Their walls are so thin that some of the blood can pass out and bathe the surrounding cells and diffusion of dissolved substances can take place between blood and tissues.

Capillaries are present all over the body in every organ so that all living cells are near to some of them. It is the capillaries that cause the skin to be pink and lean meat to look red; they are too small to see individually, but they are so close together that they have this effect.

1. Put a drop of clove or cedar wood oil on to the colourless skin at the base of a finger nail; it will make the skin more transparent. Place your finger under a binocular microscope and shine a strong light on top of it; you should be able to see some of your own capillaries. Do they form a particular pattern?

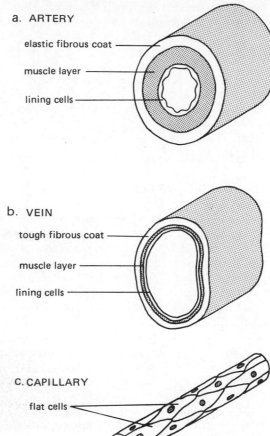

a. ARTERY

elastic fibrous coat ———
muscle layer ———
lining cells ———

b. VEIN

tough fibrous coat ———
muscle layer ———
lining cells ———

c. CAPILLARY
flat cells ———

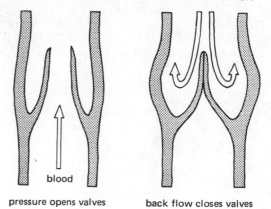

d. VEIN IN SECTION SHOWING VALVE ACTION

blood

pressure opens valves   back flow closes valves

Fig. 1:8 The structure of blood vessels.

7

2. Examine the external gills or the tail of a newt larva or a tadpole of a frog or clawed toad *(Xenopus)*. To keep the animal still while you examine it under the microscope, anaesthetise it lightly by placing it in a 0.05% solution of MS222 (tricaine methane sulphonate). Note the pulsation due to the heart action. Replace the animal in fresh water as soon as you have finished your observations. What exchange of material is taking place in these capillaries?

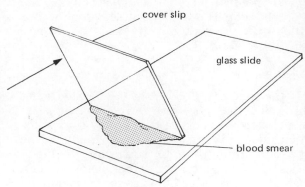

Fig. 1:9 Making a blood smear.

BLOOD

This is a liquid tissue consisting of cells or **corpuscles** which float in a pale straw-coloured fluid, the **plasma**. You can separate these components by putting a few cm^3 of blood into a tube and centrifuging it. You can study these blood corpuscles by preparing and examining a microscope preparation, using your own blood.

1. Sterilise the tip of a finger with some surgical spirit. Draw more blood into your finger by swinging your arm round, holding it downwards and wrapping a handkerchief tightly around the base of the finger.
2. Make a quick jab with a sterile lancet and put a drop of blood on to a very clean slide near one end.
3. Make a thin smear using another slide or a thick cover slip as shown in Fig. 1:9. Allow the smear to dry.
4. Pipette some Leishman's stain on to the dried smear. Leave for five minutes, then carefully wash off excess stain with distilled water.
5. Gently wave the slide in the air to dry (or place it over a bench lamp).

Examine the preparation under the high power of the microscope; there is no need to use a coverslip.

Note the very numerous red corpuscles which have no nuclei, and the white corpuscles which may be easily distinguished because their nuclei will be stained blue. Compare their size with that of the red cells. Are their nuclei all alike?

Red corpuscles (erythrocytes)

Their function is to absorb, transport and release oxygen. They are produced continu-

ously at the rate of ten million per second in the bone marrow of the ribs, sternum and long bones especially. Before being released into the blood they lose their nuclei and become flattened, bi-concave discs (Fig. 1:10). Each corpuscle consists of an elastic membrane which encloses cytoplasm and a high concentration of **haemoglobin**.

Haemoglobin, which is purplish red in colour, is a protein containing iron. Under conditions of high oxygen concentration one molecule of haemoglobin can combine loosely with four molecules of oxygen to form **oxyhaemoglobin** which is bright red. However, in conditions of low oxygen concentration the oxyhaemoglobin will give up its oxygen and revert to haemoglobin. Thus, when the blood passes through the tissues of

Fig. 1:10 Blood components.

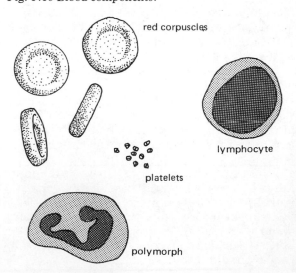

8

the lung oxygen is taken up, and when the blood reaches other tissues, oxygen is released.

LUGS
oxygen concentration in
alveoli higher than
in the blood

haemoglobin ————————→ oxyhaemoglobin
(purplish red)   (bright red)
+ oxygen ←————————

oxygen concentration in
blood higher than
in the tissues
TISSUES

The presence of haemoglobin in the blood allows 20–30 times more oxygen to be carried than if the oxygen was in simple solution.

In conditions where the amount of oxygen in the atmosphere is reduced, e.g. at high altitudes, the body responds by increasing the number of red corpuscles in the blood. This is one reason why climbers spend time acclimatising themselves at moderate altitudes before attempting the highest mountains. Without this training they would quickly become exhausted.

The red corpuscles become worn out after about four months and are broken down in the liver and spleen into bile salts and other substances which are used again to make more corpuscles.

Haemoglobin also combines with the poisonous gas carbon monoxide, which is a constituent of coal gas (not natural gas), car exhaust fumes and to a lesser extent in cigarette smoke (Book 1 p. 143). When combination occurs, **carboxyhaemoglobin** is formed:

carbon monoxide + haemoglobin → carboxyhaemoglobin
(purplish red)   (cherry pink)

This reaction is irreversible, so it follows that if too much carbon monoxide is breathed in, there is insufficient haemoglobin left to combine with oxygen, and this condition quickly leads to death.

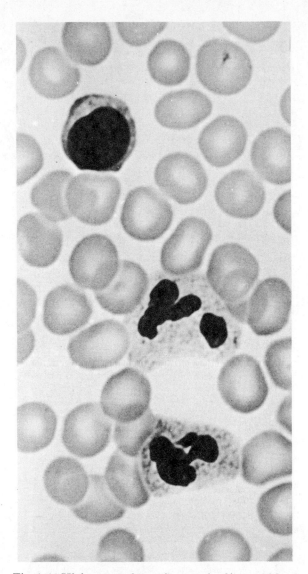

Fig. 1:11 High power photomicrograph of human blood showing a lymphocyte, two polymorphs and numerous red corpuscles.

White corpuscles (leucocytes)

In human blood these are in the proportion of one white corpuscle to approximately 500 red. They are concerned with combating infection and disease. You will probably have seen in your blood preparation two types:

1. Polymorphs
These are formed in the bone marrow; they have an irregularly shaped nucleus. They can ingest bacteria in the same manner as amoeba,

9

flowing round them and digesting them with enzymes. Polymorphs are able to squeeze through the walls of capillaries into the surrounding tissues, and if there is a wound, enormous numbers will migrate to the site to prevent the spread of bacteria into the body.

2. Lymphocytes

These have a large nucleus which nearly fills the cell. They attack bacteria by producing chemical substances, **antibodies**, which react with the surfaces of the bacteria and often cause them to clump together. The polymorphs can then ingest them. Disease-producing organisms (pathogens) liberate poisonous excretory substances or **toxins**. It is these which often cause fever. The lymphocytes produce other kinds of antibodies called **antitoxins** which neutralise the toxins, but this often takes time. Lymphocytes are found in great numbers in the lymphatic system (p. 13).

One of the big problems with organ transplants is that white blood corpuscles recognise, for example, a newly transplanted kidney as 'enemy' tissue and immediately invade and try to destroy it. To prevent this, the patient has to take various types of drugs to make his own defence mechanisms less effective. This, of course, may make him more susceptible to infection. Many of the deaths following the early heart transplants were due not so much to rejection of the 'foreign' heart but to secondary infection in other parts of the body.

Platelets

These are minute fragments budded off from certain large cells found in the bone marrow. They are concerned with the blood clotting process (p. 11).

Plasma

This is the fluid part of the blood; it contains a large number of substances in solution (see table opposite). The commonest mineral salt carried is sodium chloride. Some of the proteins cannot pass through the capillary walls as their molecules are too large, but sugars (chiefly glucose), amino acids and fatty substances can pass through by diffusion, provided there is a

THE MAIN FUNCTIONS OF THE BLOOD

TRANSPORT

1. Oxygen	Carried from lungs to tissues as oxyhaemoglobin in the red corpuscles.
2. Carbon dioxide	Taken from tissues to lungs as bicarbonate, mainly in the plasma.
3. Waste substances	Chiefly urea; formed in the liver and transported to the kidneys (p. 72).
4. Food	Most substances transported in solution in the plasma from the small intestine wall to the liver via the portal vein; then from the liver to the rest of the body. Most fatty substances enter the blood via the lymph (p. 13).
5. Hormones	Produced by ductless glands (p. 24); transported in the plasma.
6. Heat	Produced during respiration, particularly in muscles and in the liver. Distributed throughout the body, but regulated through the contraction or dilation of arterioles (p. 78).
DEFENCE AGAINST HARMFUL ORGANISMS:	a) White corpuscles (polymorphs) ingest bacteria.
	b) White corpuscles (lymphocytes) produce antibodies which neutralise harmful chemical substances.
	c) Substances in platelets and plasma cause blood to clot at a wound. Harmful bacteria are prevented from entering.

concentration gradient.

Transport of carbon dioxide

Most of the carbon dioxide is also carried in the plasma in the form of bicarbonate ions. When carbon dioxide is formed in the cells during respiration it builds up a concentration higher than that in the blood, and so the carbon dioxide passes along the diffusion gradient into the blood. Here it dissolves in the plasma and forms carbonic acid which then dissociates into hydrogen and bicarbonate ions:

$$\underset{\text{water}}{H_2O} + \underset{\substack{\text{carbon} \\ \text{dioxide}}}{CO_2} \rightarrow \underset{\substack{\text{carbonic} \\ \text{acid}}}{H_2CO_3}$$

$$\underset{\substack{\text{carbonic} \\ \text{acid}}}{H_2CO_3} \rightarrow \underset{\substack{\text{hydrogen} \\ \text{ion}}}{H^+} + \underset{\substack{\text{bicarbonate} \\ \text{ion}}}{HCO_3^-}$$

When the blood reaches the lung capillaries the reverse action takes place and the carbon dioxide diffuses out of the blood into the alveoli, once more passing along the diffusion gradient:

$$HCO_3^- + H^+ \rightarrow H_2CO_3 \rightarrow H_2O + CO_2$$

Clotting of blood

The blood plasma contains a protein, **fibrinogen**. When tissues are damaged and bleeding occurs, substances in the blood platelets and plasma are released and 'trigger' off the clotting mechanism. Basically, clotting involves the conversion of fibrinogen into **fibrin** which is precipitated from the plasma in the form of a network of fibres in which blood corpuscles become enmeshed. Thus a clot is formed over the wound and the bleeding is stopped. Later the clot hardens to form a scab which protects the underlying tissues from bacterial infection. When new skin has been formed underneath, the scab loosens and comes off.

COMPOSITION OF THE BLOOD

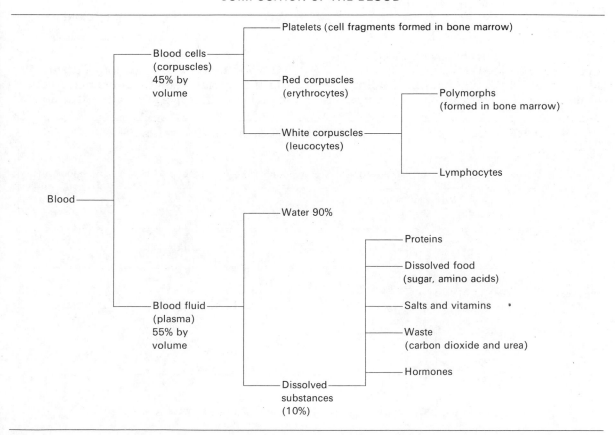

11

Clots can also occur without any external damage to the tissues; this may be due to a diseased artery wall. Portions of the clot may be carried away into the blood stream and block a smaller vessel, so stopping the circulation to the organ concerned. If this happens in the coronary artery which supplies the heart, a heart attack or **coronary thrombosis** results. If it occurs in a brain artery, it causes a stroke or **cerebral thrombosis**. A cerebral thrombosis can also be caused by the bursting of an artery in the brain. This is more usual in elderly people who have diseased arteries and a high blood pressure. A hereditary disease, **haemophilia**, is a condition where the blood is unable to clot properly, so wounds will not heal. In consequence a person with this disease may suffer a serious loss of blood through even a minor wound, or experience very severe pain and possible damage to the tissues through the pressure of blood collecting at the site of an internal injury, such as a bad bruise.

The composition of blood and its functions are summarised in the tables on pp. 10 and 11.

Blood groups

A person suffering from severe bleeding may have to have a blood transfusion, but first his blood group must be known. Mixing the wrong types of blood can be very dangerous and often fatal, as this causes the red corpuscles to clump together to form clots large enough to block a blood vessel. Blood types which do not mix are said to be **incompatible**. Two main types of protein called **antigens** occur on the surface of red blood cells. They are called A and B. People may have either of these, both, or none at all; so the four groups are known as A, B, AB and O. Similarly, the plasma may contain certain antibodies known as anti-A and anti-B. As these antibodies react with their corresponding antigen causing clumping, normal blood cannot contain, for example, antigen A on the red cells and its antibody, anti-A in the plasma. The details are given in the following table:

Group	Antigen on red blood cell	Antibody in plasma
A	A	anti-B
B	B	anti-A
AB	A and B	NONE
O	NONE	anti-A and anti-B

The blood used for transfusion contains very little of the original plasma obtained from the donor, hence it is the corpuscles of the donor's

		Antigens on donor's red blood cells			
		A (Group A)	B (Group B)	A and B (Group AB)	NONE (Group O)
Antibodies in receiver's plasma	anti-B (Group A)	✓	CLUMP	CLUMP	✓
	anti-A (Group B)	CLUMP	✓	CLUMP	✓
	NONE (Group AB)	✓	✓	✓	✓
	anti-A and anti-B (Group O)	CLUMP	CLUMP	CLUMP	✓

✓ = will mix

In Britain about 45–50% of people belong to Group O and about 40% to Group A.

blood that must be compatible with the plasma of the recipient. So, if Group A blood was transfused into a Group B recipient, the anti-A antibody in the receiver's plasma would immediately react with the A factor in the donor's red cells and cause clumping.

AB blood contains no antibodies and so will not clot the red cells of any group added to it; so people in this category are called **universal recipients**. Those of the O group are called **universal donors** as their red cells contain no antigens. This is summarised in the table opposite.

The National Blood Transfusion Service is dependent upon blood donated voluntarily; it is stored in 'blood banks' ready for any emergency. The donors give a pint of blood each time. New donors between the ages of 18 and 65 are always needed.

The rhesus factor

There is another antigen of red blood cells which is present in 85% of the people of Britain; this is known as the **rhesus factor**. People who have this are said to be rhesus positive (Rh +). Those who do not have this factor are termed rhesus negative (Rh −). Normally they do not carry in their plasma an antibody to this factor. However, if Rh + blood is transfused into the blood of a Rh − person, antibodies will be formed and these are capable of destroying Rh + red cells. Under certain circumstances this is a potential hazard for babies.

If a Rh + man marries a Rh − woman, some of the children are likely to be Rh +. At birth there is always some mixing of blood between the circulations of mother and baby and this may occasionally happen during pregnancy. So, if a child is Rh + some of its blood will leak into its mother's circulation and cause antibodies to form in her blood. If the mother has more children, the amount of antibody in her blood will increase with each pregnancy, and in some instances the antibodies in her blood may pass into the baby's blood in sufficient quantities to produce very serious anaemia and even death. Fortunately these cases are infrequent, and when they do occur, the baby is given a complete transfusion soon after birth so that its blood is replaced by blood containing no antibodies to the rhesus factor.

It is now possible for this transfusion to be carried out before birth. Another recently developed technique is for the mother to be given an injection shortly after the birth of her first child which neutralises the effect of the harmful antibody.

The lymphatic system

If we lightly graze our skin a pale straw-coloured fluid oozes out; this is **lymph**. Much of it is derived from the fluid which seeps out through the blood capillary walls and bathes all our tissues. Lymph is the vital link between blood and tissues by which essential substances pass from blood to cells and excretory products from cells to blood (Fig. 1:12). It is similar in constitution to plasma, except that some of the plasma proteins are absent. Lymph contains no red blood cells, but considerably more lymphocytes than are present in blood.

Some of the lymph finds its way back into the blood stream, but most of it enters a network of blind-ending lymph capillaries. These lead to larger lymph vessels, rather like thin-walled veins. The surrounding muscles squeeze on these vessels and help to push the lymph along; one-way valves assist this process. At certain places the lymph passes into lymph nodes.

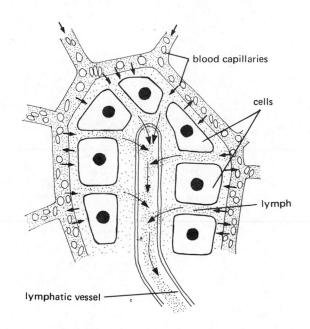

Fig. 1:12 Diagram showing the relationship between blood capillaries and the lymphatic system.

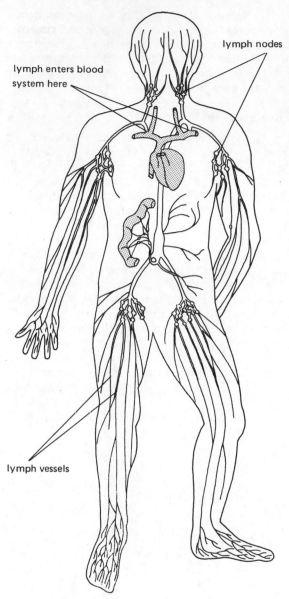

lymph enters blood system here

lymph nodes

lymph vessels

These act as filters and poisonous substances are inactivated here. Lymphocytes are formed in these nodes and when we suffer from an infection the nodes swell up. For example, when we have a badly infected throat the glands on either side of the throat usually swell.

The tonsils which lie at the back of the throat, and the adenoids at the juncture between the back of the nose and the throat, are also made of lymphatic tissue. When we have tonsilitis the tonsils swell up in the process of combating the infection, causing them to become painful.

The lymph, having been 'purified' by the lymph nodes, finally enters the blood stream via two large veins on either side of the neck. The general distribution of the lymph channels and the main lymph nodes are shown in Fig. 1:13.

You will recall that the lymphatic system is also largely responsible for absorbing fatty substances from the intestine, the lacteals in the villi being part of this system.

Lymph, therefore, has both a circulatory and a defence function. It keeps the 'chemical environment' around the tissues constant. It is perhaps surprising that there is four to five times as much lymph in the body as there is blood.

Fig. 1:13 The lymphatic system in man, much simplified.

2

Transporting materials within the plant

You may remember (Book 1 Ch. 5) that in the root the xylem tissue was situated towards the centre while in the stem it was arranged in bundles near the outside. If the distribution of dye in your sections corresponds with that of the xylem it would confirm that this tissue in concerned with water conduction. The xylem consists of long vessels with thickened walls of lignin (wood) which form continuous tubes through the root, stem and leaves.

What happens to this water? Does it remain in the plant or is it given out again into the atmosphere? You could see if it passes out again by using the apparatus in Figure 2:1, because if the atmosphere inside the bell jar became saturated with water vapour it would condense as drops on the cold walls of the jar.

Is there anything in plants which corresponds to a blood system?

Every part of a plant needs water and if we grow plants in the house we have to water them regularly if they are to survive. It is therefore reasonable to suppose that there is a means of absorbing and transporting water from the roots to all the other parts. To test this hypothesis you can colour the water with a dye; if water is conducted the dye should stain the tissues in any region through which the water passes.

Collect several small plants with their roots attached, wash away any soil which clings to them and place them in a beaker of 1% eosin so that the roots are covered by this red dye. Place in the same beaker some shoots from other plants which bear white flowers.

After half an hour, take out one of the complete plants and cut it across at 10mm intervals starting from the root end. Examine the cut cnds with a hand lens and look out for traces of red.

If you see any red colour, is the dye evenly distributed over the whole surface or is it confined to certain areas? How far has the dye travelled up the plant?

Leave the other plants in the dye for twenty-four hours and then examine the stems, leaves and flowers. What deductions can you make from your observations regarding a possible transport system?

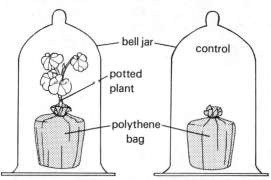

Fig. 2:1 Apparatus to determine whether water is given out by a green plant.

Set up the apparatus as in Fig. 2:1. Why should you water the plant first? Why should you enclose the soil in a polythene bag? Why should there be a control containing no plant?

Leave the apparatus for a few hours, or overnight.

If the walls of the bell jar become moist, test whether this is water using white anhydrous copper sulphate; it will turn blue in the presence of water.

From these experiments you should have deduced that a plant has a transport system for water. We now have to consider the mechanics of the process.

How is water absorbed?

Examine some mustard seedlings which have been grown on wet filter paper. Note the mass of fine threads coming from the radicle. These are root hairs through which water enters the plant. Gently squash a portion of the radicle between slide and coverslip in a drop of water and examine under a microscope. Are these root hairs extensions of the outer cells of the roots? Note the thinness of the walls of the root hairs.

It is not completely understood how the water enters the root hairs and passes inwards from cell to cell until it gets into the xylem vessels, but there is no doubt that osmosis plays an important part.

We saw in Book 1 Ch. 12 that every living cell acts as an osmotic system, the cytoplasm lining the cell wall being the semi-permeable membrane. If you look at Fig. 2:2 you will see that the root hairs grow out into the spaces between the soil particles and that the hairs are surrounded by moisture. This soil water is an extremely dilute solution of salts—more dilute than that of the cell sap in the root hair; water will therefore pass into the vacuole of the root hair by osmosis. The entry of water dilutes the contents of the root hair vacuole so that it becomes weaker than its neighbour. Therefore water passes into the neighbouring cell which in turn becomes diluted, causing water to pass yet further in and so on until finally water enters the xylem vessels. As there are vast numbers of root hairs and root cells involved, a pressure in the xylem vessels develops which forces the water upwards. This total pressure is known as **root pressure**.

Root pressure is not the *main* cause of movement of water in the xylem as we shall see later, but it is certainly one factor. Root pressure may be demonstrated on a vigorously growing potted plant such as a geranium (Fig. 2:3). The stem is cut near ground level and connected to a glass tube by means of strong rubber tubing, the joints being bound tightly. Water is added until it can be seen above the rubber tube; the level of the water is then marked. If there is a pressure of water from below, the level will rise.

In spring and early summer, root pressure

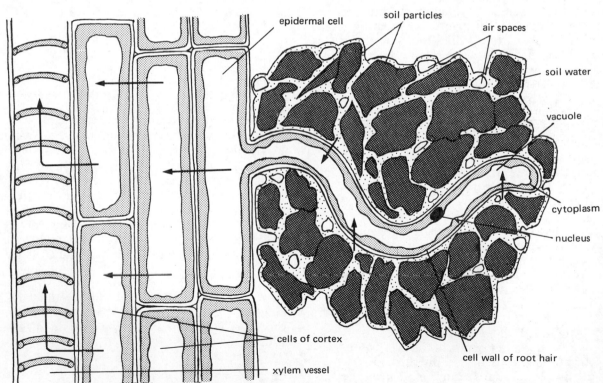

Fig. 2:2 Simplified diagram of part of a plant root in longitudinal section showing the relationship of the root hair to the soil water. Arrows indicate the movement of water.

16

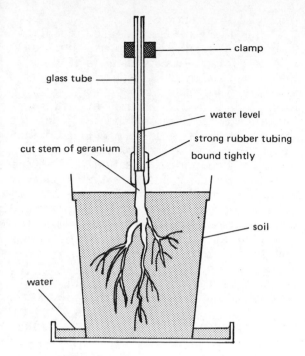

Fig. 2:3 Apparatus for demonstrating root pressure.

in trees can be quite considerable and if a branch is cut at this time, sap oozes from the cut surface, but at other times the pressure is greatly reduced and plays little part in moving the water up the xylem.

How does the plant lose water?

We saw earlier that water evaporated from the surface of the shoot and it became visible as droplets when it condensed on the cold walls of the bell jar. This evaporation of water from plants is called **transpiration**. It occurs through the surface of the whole shoot, but some parts lose more water than others. You can investigate this by comparing the evaporation rate from the two surfaces of a leaf. The principle of the experiment is to use small pieces of cobalt chloride paper on both sides of a leaf and see if one turns from blue (dry) to pink (damp) more quickly than the other.

Use two rather different leaves such as sycamore and iris. Make sure that the pieces of cobalt chloride paper you use have been kept in a desiccator to remove all traces of moisture. Why should you handle them with forceps and not with your fingers?

Place a piece of cobalt chloride paper on one of the leaves and cover it with Sellotape, being careful to make an air-tight seal (Fig. 2:4). Turn the leaf over and repeat the procedure. Treat the second leaf in the same way.

Which sides of the two leaves lose water most quickly? (The underside of the iris is the one where the mid-rib is most prominent.)

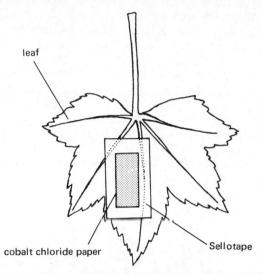

Fig. 2:4 Method of comparing the rates of transpiration from the two sides of a leaf.

It is reasonable to suppose that this result could be related to the distribution of the stomata on the leaves. Are there more stomata on one surface than the other in the two specimens? You can find out in this way:

Paint some nail varnish on to a part of each of the two surfaces of each leaf. When dry, carefully peel off each film and examine it under the microscope. The impression of each stoma will show up. Compare the number of stomata on the two surfaces of both leaves by counting all the impressions you can see under the high power field of the microscope. Choose five places for each surface and take the average, ignoring the stomata which are more than half out of the field of view.

How do these results compare with those from your previous experiment?

From your observations you should have seen that leaves differ in the number and distribution of their stomata. In some leaves

17

there may be as many as 20,000 stomata per cm² on the under surface and as many as 10,000 on the upper, but many leaves have no stomata at all on the upper surface.

Leaves also differ in the thickness of their cuticle. This is a hard, waxy secretion which covers the outer walls of the epidermal cells. In leaves such as holly it is very thick and makes the leaf stiff; in flimsy leaves it is thin. Some water can evaporate through a very thin cuticle, but when thick, the cuticle is impervious to water. If there were no stomata, all water loss could be prevented by having a thick cuticle, but the plant has to have some means of taking in and giving out gases such as oxygen and carbon dioxide, hence stomata are needed and some water will automatically evaporate too. Transpiration is therefore an inevitable process in plants.

What adaptations would you expect to find regarding the distribution of stomata and the thickness of the cuticle in:
1. Plants living in damp, shaded woodland?
2. Plants living in dry, exposed situations?

Transpiration has its dangers if there is insufficient water in the soil to replace what is lost. If you forget to water potted plants, the leaves soon wilt. This occurs when transpiration exceeds absorption, and the cells lose their turgidity. If wilting is prolonged, it may cause the death of some plants. Wild plants living in their natural situations are seldom seen to be wilted, although under extreme conditions this may happen. This suggests that for most plants the water absorbed is equal to the water transpired.

You could find out if this is true by using a **weight potometer**.

Set up the apparatus as in Fig. 2:6.

Measure the change in weight of the whole apparatus over a period of 24 hours. This will give you the weight of water lost to the atmosphere.

Calculate the amount of water absorbed by the plant over the same period by noting the change in volume of water in the flask. Do this by finding out how much water needs to be added to bring the water level up to the original mark.

As 1cm³ of water weighs 1g, check if the weight loss in grammes equals the volume in cm³ of water added.

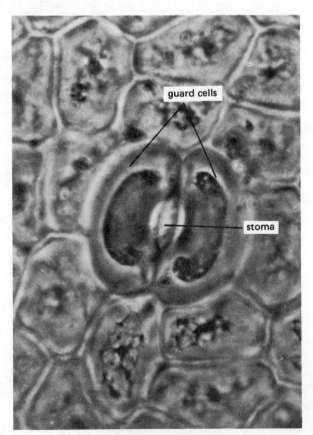

Fig. 2:5 High power photomicrograph of a stoma and guard cells in surface view (from a box leaf).

guard cells

stoma

The rate of transpiration

Transpiration is essentially the evaporation of water from the plant shoot—mainly from the leaves. The evaporation actually takes place *inside* the leaves from the wet surfaces of the cells into the air spaces; it then diffuses out, mainly through the stomata. This diffusion process will depend on the conditions of the atmosphere outside, such as the humidity, temperature, the amount of wind and perhaps light. You can test the effect of these conditions on the rate of transpiration by using a **bubble potometer** (Fig. 2:7) which is more sensitive than the weight potometer and gives continuous readings. This potometer really measures the rate of *absorption*, but as shown in the last experiment, when plants are given plenty of water, absorption is, for all practical purposes,

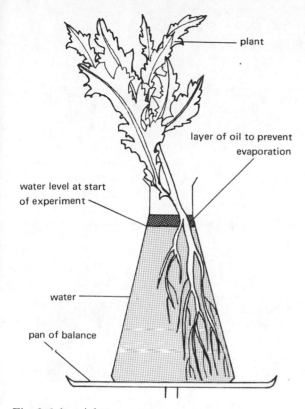

plant

layer of oil to prevent
evaporation

water level at start
of experiment

water

pan of balance

Fig. 2:6 A weight potometer.

replace. Take the time for the bubble to move between the two chosen points, then, with the tube under water, squeeze the plastic tube to expel the bubble, otherwise air will collect below the cut end of the shoot. Another bubble can then be introduced. Using this apparatus, devise a means for comparing the rates of transpiration under different conditions. Give the shoot time to become adjusted to the change before taking readings. Here are some suggestions: an electric fan is a good, steady source of wind, and plastic hoods—one black and one transparent—could be used to cover the shoot loosely to show the effect of bright and dim light. What about the effect of temperature or humidity? Be critical of your methods to ensure that by altering *one* condition you do not alter another as well.

What effect on the rate of transpiration would you expect if you covered a) the upper surface or b) the lower surface of each leaf of the shoot with Vaseline? Would any changes

equal to transpiration, so this apparatus is used to measure transpiration.

Cut a shoot round enough and with a diameter large enough to make an airtight joint with the plastic tube. Plunge it immediately into a jar of water to prevent air from entering the xylem vessels. Assemble the apparatus under water to ensure that no air gets in for the same reason, taking care not to wet the leaves (why?). Leave the apparatus for a few minutes to allow the plant to become adjusted to the external conditions.

The principle is that if a bubble of air is introduced into the capillary tube it will be drawn up by the water column as absorption of water takes place. The rate of movement of the bubble can be measured by taking the time for it to travel between two points marked on the capillary tube or on a scale fixed parallel to it. (An average of three readings is better than one—why?)

To introduce the bubble, lift the tube out of the beaker, blot the end with filter paper and

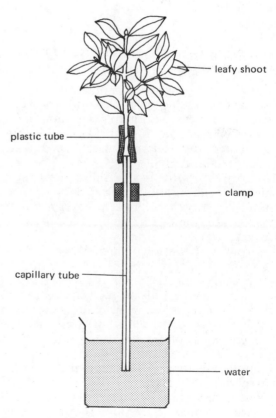

leafy shoot

plastic tube

clamp

capillary tube

water

Fig. 2:7 A simple form of bubble potometer.

be dependent upon the species of leaf being used? What change in rate would you expect if you cut off half the leaves with scissors?

We can summarise the effect of external conditions on the rate of transpiration as follows:

1. An increase in temperature increases the rate because a) air can contain more water vapour at higher temperatures, so it increases the diffusion gradient between the air in the leaf and the air outside, b) it speeds up the process of evaporation.
2. An increase in the humidity of the air decreases the rate because the diffusion gradient is lessened, and if the air is saturated, transpiration stops altogether.
3. Wind increases the rate, because as soon as the air outside the leaf receives moisture from inside by diffusion, it is replaced by drier air, thus keeping the gradient higher.
4. Light. This has no *direct* effect, but light does cause stomata to open more widely (Book 1 p. 153) and this may cause an increase in the rate of transpiration.

The mechanism by which the water travels through the plant

We have seen that there is a push from below due to root pressure on the columns of water in the xylem vessels, but this is seldom large and at some seasons it is nil. How does the water reach the top of a tree like a giant redwood 120m high?

The principles involved can be demonstrated by using the apparatus in Fig. 2:8:

a) is a leafy shoot immersed in water in a tube dipping into some immiscible fluid such as mercury.

b) is a porous pot fitted up in the same way.

The wall of a porous pot is a mass of minute 'tubes' and when the water evaporates at the surface it causes a pull on the water columns and the mercury is drawn up. Similarly, when the leaves transpire there is a pulling effect on the continuous columns of water in the xylem vessels. The top ends of these vessels are surrounded by the leaf's mesophyll cells which contain sap, so the water is continuous from the xylem vessels to the walls of the mesophyll cells from which it evaporates into the air spaces causing the pull. The water column

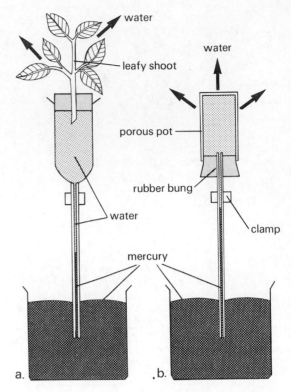

Fig. 2:8 Apparatus for investigating the effect on a column of water of a) transpiration b) evaporation.

does not break because of its great tensile strength. This is a property of water you demonstrate every time you suck up a fluid through a straw.

We now have a picture of the water-conducting system of a tree. Water is absorbed by osmosis from the soil by the root hairs and is passed into the xylem vessels which form a continuous system of tubes through root and stem into the leaves; here the water evaporates and passes into the atmosphere. The evaporation creates the main pull from above, root pressure gives a variable and minor push from below. The result is a continuous column of moving water, the **transpiration stream**.

The amount of water passing through a plant is often considerable. For example, it was estimated that the Egyptian cotton crop transpired 1·2 tonnes of water per sq. dekametre per day (1·7 litres per plant). It follows therefore that areas of forest significantly affect the degree of saturation of the air above them, so that when air currents bring air which is already nearly saturated to a forest area, it be-

20

comes fully saturated and comes down as rain; this is why forest areas often have a higher rainfall than areas nearby.

Transport of mineral salts

You will recall that mineral salts are necessary for plant nutrition and that they are obtained from the soil in solution through the root hairs. The salts are in the form of electrically charged ions. Thus sodium chloride (NaCl) is in the form of Na^+ and Cl^-, and magnesium sulphate ($MgSO_4$) occurs as Mg^{2+} and SO_4^{2-}. They are *not* absorbed into the root hairs by the simple process of diffusion, as you would expect, for the cell sap of the root hairs contains a higher concentration of ions than in the soil water under normal conditions, and if diffusion did occur, the ions would pass from the plant to the soil. The detailed mechanism of absorption is not fully understood, but it must involve the use of energy by the cytoplasm, because if roots are deprived of oxygen their ability to absorb ions is reduced. It has been found that cytoplasm, when surrounded by solutions of ions, absorbs some ions more than others and in this sense cytoplasm can be said to be selective.

Once absorbed, the ions travel in the water in the xylem vessels and pass to the growing points of the plants where they are used for growth.

Transport of manufactured food

Food such as sugar is synthesised in the green parts of plants, mainly the leaves, but this food has to be transported to all the living cells, especially those which are actively growing and those which store food.

The veins of a leaf consist of xylem and phloem, and these tissues are continuous with those of the stem. The following experiments provide evidence that food is transported in the phloem cells.

Phloem sieve tubes (Book 1, p. 53) are extremely small and the analysis of their contents is not easy, but with the help of aphids (greenfly) this has been done. When you see aphids clustering round the young stems of roses or broad bean plants they are feeding on the plant juices. To obtain this juice an aphid pierces the plant tissues with its long proboscis.

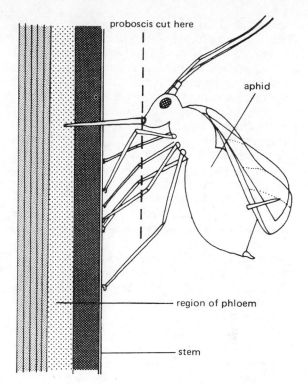

Fig. 2:9 Method of using the proboscis of an aphid for extracting fluid from phloem sieve tubes.

It can be shown that when a feeding aphid is killed and the stem carefully sectioned, the proboscis only penetrates as far as a phloem sieve tube. This proboscis also provides a ready-made means of obtaining the juice for analysis! The experiment can be done in this way. An aphid is killed while in the act of feeding and the body is then carefully cut away, leaving the hollow proboscis still inserted into the phloem (Fig. 2:9). It is found that because the contents of the phloem sieve tubes are under slight pressure the fluid slowly exudes from the cut end of the proboscis in the form of drops; these drops are then collected and analysed. The fluid is found to contain sugars and amino acids.

Not surprisingly, aphids absorb so much sugar from the phloem that they cannot assimilate all of it and it passes out of the anus as a sticky syrup called **honey-dew**. Leaves which have been attacked by aphids often feel sticky as a result.

Further experiments to illustrate the conduction of sugars by the phloem have been done by removing a ring of bark from a shoot to expose the wood. This in effect removes all tissues from the cambium outwards, including the phloem. After a few days, when the tissues above and below the ring were analysed it was shown that food had accumulated above the ring, but was not present below it. If left for some time, the stem increased in thickness immediately above the ring, but no growth occurred below it (Fig. 2:11). So any damage to the phloem all around the stem will prevent food from passing down to the roots and the tree will eventually die. This is a fact of great economic importance because certain mammals gnaw the bark of trees to get at the food stored in the phloem, especially during hard winters when food is scarce. Voles do this to young saplings at ground level and rabbits can do much damage to older ones. Foresters find it economically worth while to enclose new plantations with wire netting to prevent rabbits from entering.

Foresters also encourage predators such as foxes, badgers, hawks and owls as they help to keep down the population of voles and rabbits. Grey squirrels too do great damage, particularly to beech and sycamore, and for this reason, in some parts it is impossible to grow these trees as a crop. When you next go into a wood look out for evidence of bark having been gnawed off saplings and trees. Note the species of tree, the

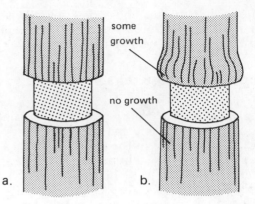

Fig. 2:11 Ringing experiment: a) ring of bark removed b) the same stem after some weeks.

position of the damage, whether the damage is recent or old, and the size of tooth marks if these are visible. From these observations you could find out which species had caused the damage. Also look out for the effect of such damage on the tree as a whole.

Fig. 2:10 One of these photographs shows typical rabbit damage, the other damage by grey squirrels. Can you determine which is which?

3

Reproduction and growth

Reproduction is a process which takes place in all living organisms. It is the ability to produce new individuals with the same general characteristics as the parents. Reproduction ensures that life is perpetuated.

In this and the next two chapters we will be concerned mainly with reproduction and growth in man and flowering plants, but first let us summarise the principal methods of reproduction in organisms generally.

There are two main methods, **sexual** and **asexual**. When sexual reproduction occurs the new individuals are formed as a result of the fusion of two special nuclei from different cells; when asexual reproduction takes place the new individuals are not formed in this manner.

Asexual reproduction

There are many ways in which this is brought about, some of which we have already considered:

1. Binary fission. This occurs when organisms such as amoeba and bacteria divide into two more or less equal parts. It is a quick and efficient method for simple organisms when conditions are favourable.
2. Spore formation. This occurs when an organism produces special structures called spores which are capable of growing into new organisms without the fusion of any nuclei. Many of the lower plants—for example, fungi, bacteria, mosses, liverworts and ferns—use this method; also some animals, such as the malarial parasite when it multiplies within the red blood corpuscles. Spores are usually produced in vast numbers, so theoretically a great many new individuals may result. In practice the majority seldom survive.
3. Budding. This occurs when a new individual gradually grows from the parent and eventually separates. We have seen this in *Hydra*, but sea anemones occasionally use the same method, the buds being formed internally. Yeast is a fungus which also produces new individuals by budding.
4. Vegetative reproduction. This term is used for a special type of budding much used by flowering plants; the buds grow out from the parent stem and form shoots which eventually separate into new plants. Propagation by runners, suckers, corms, bulbs, tubers and rhizomes are examples (p. 37).
5. Parthenogenesis. This method is unusual. It occurs when an egg cell remains unfertilized, but nevertheless develops into a new individual. It takes place in drone bees, aphids, water fleas *(Daphnia)* and the dandelion (p. 49).

Sexual reproduction

This results from the fusion of nuclei from separate sex cells called **gametes**, the fusion process being called **fertilization**. In all but the simplest organisms such as *Spirogyra*, the gametes are basically of two kinds: sperms which are derived from the male and ova (eggs) from the female. Sperms are motile and swim to the ovum to fertilize it. In flowering plants, however, the male gametes are formed in the pollen and are not motile. Fertilization may take place outside the body of the parent as in many fish and amphibians (**external fertilization**), or inside the female as in insects and mammals (**internal fertilization**). In animals the gametes are formed in special reproductive organs called **gonads**—testes in the male and ovaries in the female. When the gametes fuse a **zygote** is formed which grows by cell division into a new individual. (In some animals such as *Hydra*, earthworms and snails, both gonads are present in the same individual; they are said to be **hermaphrodite**.) The basic process of sexual reproduction can be summarised as follows:

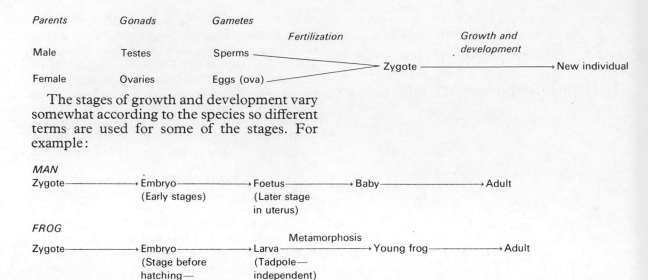

Parents	Gonads	Gametes			
			Fertilization		*Growth and development*
Male	Testes	Sperms			
				Zygote	New individual
Female	Ovaries	Eggs (ova)			

The stages of growth and development vary somewhat according to the species so different terms are used for some of the stages. For example:

MAN

Zygote ⟶ Embryo ⟶ Foetus ⟶ Baby ⟶ Adult
(Early stages) (Later stage in uterus)

FROG

Metamorphosis

Zygote ⟶ Embryo ⟶ Larva ⟶ Young frog ⟶ Adult
(Stage before hatching— dependent on egg yolk) (Tadpole— independent)

BUTTERFLY

Metamorphosis

Zygote ⟶ Embryo ⟶ Larva ⟶ Pupa ⟶ Adult
(Within egg shell) (Caterpillar) (Chrysalis) (Imago or butterfly)

Growth

Growth is an inevitable consequence of reproduction. It is the permanent increase in size of an organism due to the formation of new protoplasm. All but the simplest organisms increase in complexity when they grow. For example, each one of us has grown from a single fertilized cell.

Our growth is influenced by several factors. Heredity is important in laying down the essential features of shape and size; for this reason tall parents, whose parents are also tall, tend to have tall children (Ch. 14). However, within the general limits laid down by heredity, growth will also be affected by diet. If the diet is adequate for all needs, optimum growth for that person should occur. It is probable that the increase in average height of people in the more developed countries over the last 100 years has been due to their having had a more adequate diet.

In addition to these factors there are chemical substances called **hormones** which greatly influence growth—as well as reproduction—so we will now briefly examine the important system in our body which produces these hormones.

The endocrine system

The term **endocrine** means 'internally secreting'. This system consists of all the glands in the body which do *not* pass their secretions down ducts, but pour them directly into the capillaries of the blood system; they are therefore called **ductless glands**. The hormones which these glands secrete are carried by the blood to all parts and have a great influence on the body generally and, in some cases, on certain organs in particular. These special structures which are influenced by hormones are called **target organs**.

Not all hormones affect growth and reproduction. Some control particular metabolic processes such as the build-up and break-down of sugar (p. 75), others are concerned with co-ordinating the action of certain glands or organs (p. 110). Figure 3:1 shows the positions of the main ductless glands of the body which comprise the endocrine system. In this chapter we will consider only those glands which are more directly concerned with growth and reproduction, that is the thyroid, the pituitary and the gonads.

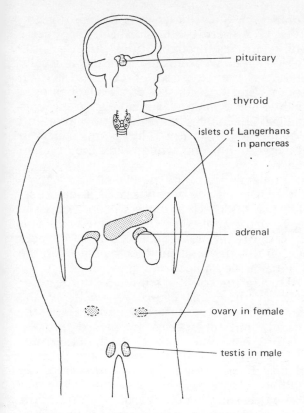

pituitary

thyroid

islets of Langerhans
in pancreas

adrenal

ovary in female

testis in male

Fig. 3:1 The main glands of the endocrine system.

The thyroid gland

This is situated in the neck just below the larynx; it consists of two lobes lying on either side of the wind pipe which are connected together. It secretes the hormone **thyroxin**, a substance containing iodine. This explains the importance of having enough iodine in the diet to allow for its synthesis (Book 1 p. 162).

Thyroxin has two main functions: it influences the rate of certain basic metabolic reactions, especially the liberation of energy from glucose in respiration, and it controls the rate of growth and so affects development.

If too little thyroxin is secreted during infancy, the child becomes a **cretin**. Cretinism is a condition where growth is stunted, sexual maturity is not reached and the person is mentally retarded. It can be cured through early treatment with thyroxin. Deficiency of thyroxin in adults results in **myxoedema**, a condition where the person becomes sluggish, fat and slow-witted.

Over-activity of the thyroid is accompanied by a swelling of the gland so that the neck enlarges; the person becomes over-active, nervous and thin, a condition known as **thyrotoxicosis**. It can be treated surgically by the removal of some of the gland, or medically by the injection of a drug which counteracts the production of the hormone.

In amphibians, thyroxin is essential for metamorphosis, although growth continues without it. Tadpoles fail to turn into frogs when the thyroid gland is destroyed, but continue to grow into abnormally large tadpoles. If small concentrations of thyroxin are added to water containing frog tadpoles at the time when they are just beginning to form hind legs, they will metamorphose into frogs in a much shorter time than those kept as a control without such treatment. As a result, the young frogs are much smaller than normal as they have not had so much time to grow.

The pituitary gland

This is a small, rounded gland connected by a stalk to the base of the brain. It exerts a tremendous and vital influence on the body, partly through its effect on the other ductless glands, and so it has rightly been called the 'master' gland of the endocrine system.

It has two lobes, anterior and posterior. One of the hormones produced by the posterior lobe stimulates the uterine muscles to contract at childbirth (p. 32). However, it is the hormones of the anterior lobe that have the more widespread effect. They include the following:
1. A growth hormone which has a special influence on the long bones and those of the hands and feet. When people have too much of this hormone during development they become giants; if too little, they become dwarfs.
2. A group of **trophic** hormones, i.e. those which stimulate other ductless glands to secrete their hormones; they include one which stimulates the thyroid to form thyroxin, another which stimulates the adrenal glands to secrete cortisone and another that influences gonad development and the production of gametes.

The gonads

The ovary and testis not only produce eggs and sperms, but they also act as ductless glands,

25

secreting hormones into the blood. The influence on human reproduction of these hormones and others from the pituitary will be discussed later in this chapter (p. 35).

HUMAN REPRODUCTION AND DEVELOPMENT

The male and female reproductive systems

In the male (Fig. 3:3) the primary reproductive organs are the two **testes** which produce the sperms. The testes lie in a special sac outside the abdominal cavity called the **scrotum**. In this position the temperature of the testes remains below that of the body, a condition which favours sperm production. The sperms are stored in the **epididymis**, an extremely long tube much coiled on itself which lies close to the testis. The epididymis leads to a long muscular tube, the **vas deferens** (pl. vasa deferentia). There are two of these, one from each testis, which unite at a point where they enter the **urethra**, just below the bladder. The urethra passes through the **penis**, the organ which transmits the sperms into the body of the female. The seminal vesicles and prostate gland secrete a fluid in which the sperms can be transferred during sexual intercourse.

In the female (Fig. 3:4), the two **ovaries** which produce the eggs are attached to the wall of the body cavity just below the kidneys. The two **fallopian tubes** carry the eggs from the ovaries to the **uterus** or womb. The latter is a muscular organ in which the foetus develops. In the non-pregnant state it is pear-shaped and quite small; its narrower end, the **cervix**, leads to the **vagina**. The vagina is muscular and connects the uterus with the outside world. It has in consequence two important functions; to receive the penis of the male during intercourse and to allow the baby to pass out of the body of the mother when it is ready to be born.

Sperm production, transmission and fertilization

Each testis consists of many convoluted, microscopic tubules bound together by connective tissue and is covered by a protective capsule to form a compact ovoid body (Fig. 3:2). The tubules produce the sperms by cell division (Fig. 3:7). Sperm production begins at puberty and continues throughout life. Puberty is the period of change from child to adult; it usually starts between the ages of 11 and 13. Each sperm (Fig. 3:8), consists of a head, composed mainly of the nucleus, and a long cytoplasmic tail which enables it to swim towards, and possibly reach, an egg. The sperms collect in the epididymis and while there, they remain inactive.

Transmission of the sperms into the body of the female occurs during sexual intercourse. The penis is largely composed of spongy tissue and during intercourse it becomes stiff and erect due to the passage of blood into the spaces within this tissue; it can then be placed in the vagina of the female.

During intercourse, stimulation of sensory cells in the penis causes the walls of the epididymis and vas deferens to contract rhythmically and the sperms are passed into the urethra; here they mix with the fluid **(semen)** secreted by the seminal vesicles and prostate gland. This fluid, together with some 500 million sperms, is then ejaculated into the vagina; it supplies the sperms with nutrients and stimulates their swimming movements. A proportion of these sperms, by lashing their tails, will swim through the cervix and uterus into the fallopian tubes. If an egg is present in

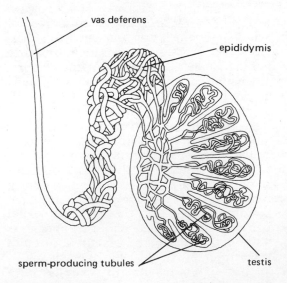

Fig. 3:2 Simplified diagram of a longitudinal section of a testis showing the arrangement of the tubules.

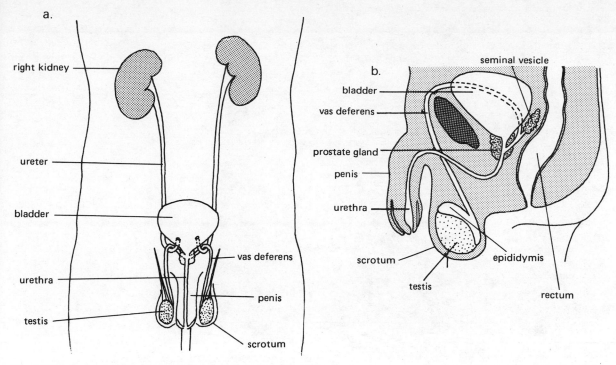

Fig. 3:3 Human reproductive system. Male: a) front b) side.

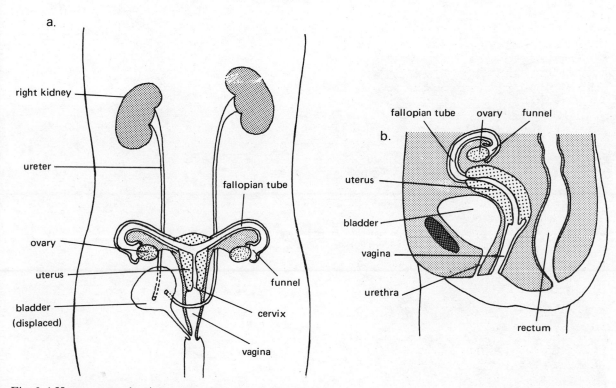

Fig. 3:4 Human reproductive system. Female: a) front b) side.

27

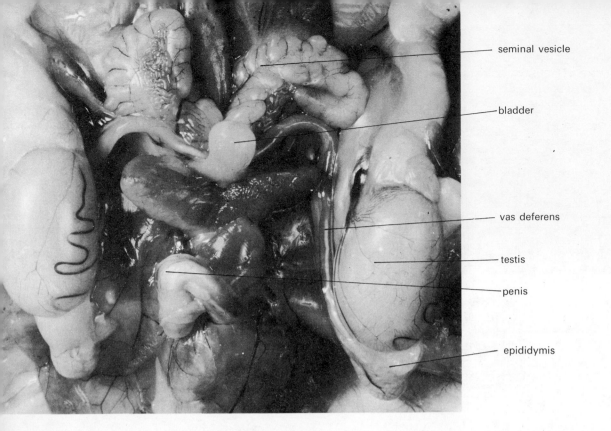

— seminal vesicle

— bladder

— vas deferens

— testis

— penis

— epididymis

Fig. 3:5 (above) Reproductive system of male rat.
Fig. 3:6 (below) Reproductive system of female rat.

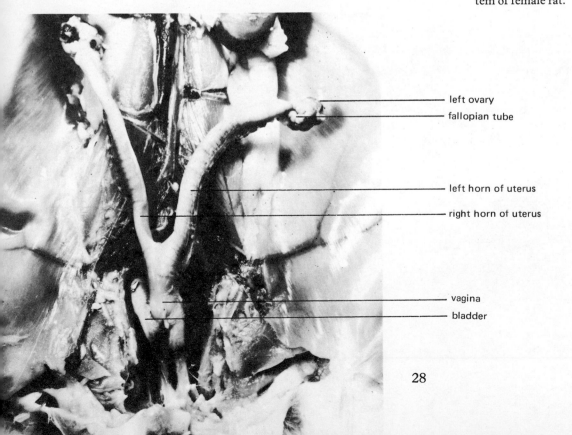

— left ovary
— fallopian tube

— left horn of uterus

— right horn of uterus

— vagina
— bladder

28

the tube, some of the sperms are likely to make contact with it, but only one actually penetrates, its nucleus fusing with that of the egg (Fig. 3:12). The other sperms are at once repelled. Usually only one egg is present in the tube, but occasionally there may be two; if both of these are fertilized, **non-identical twins** will develop.

The ovary and ovulation

At birth the ovaries of a baby girl are already formed and contain many thousands of potential egg cells, but only about 500 of these will become mature during the period of reproductive life. It is not until puberty that these potential egg cells start to form mature eggs and from this time until the age of about 50 one or other ovary produces a single egg about every 28 days. Fig. 3:11 shows a diagram of a section of an ovary to show how the potential egg cell matures, surrounded by a group of actively dividing cells; in this way a **Graafian follicle** is formed. When ripe the follicle contains a fluid-filled cavity and projects from the wall of the ovary as a small bump. **Ovulation** then

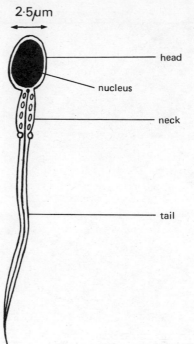

Fig. 3:8 Diagram of a human sperm.

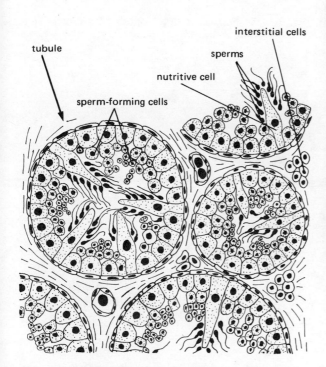

Fig. 3:7 Diagram of a section through testis tubules, much enlarged.

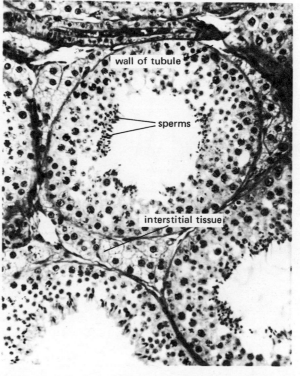

Fig. 3:9 Photomicrograph of a section through part of a testis.

29

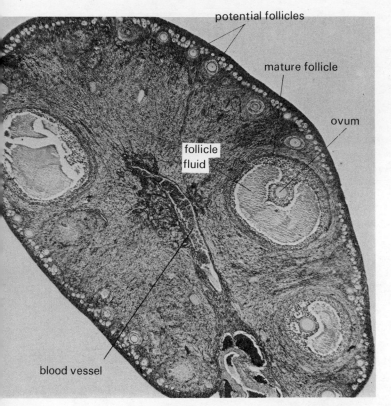

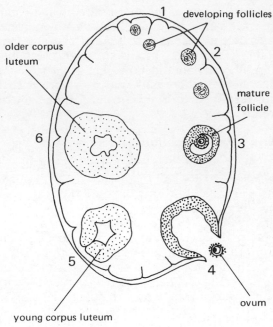

Fig. 3:10 Photomicrograph of a section through the ovary of a rabbit.

Fig. 3:11 Stages in the development of a follicle and corpus luteum in the ovary.

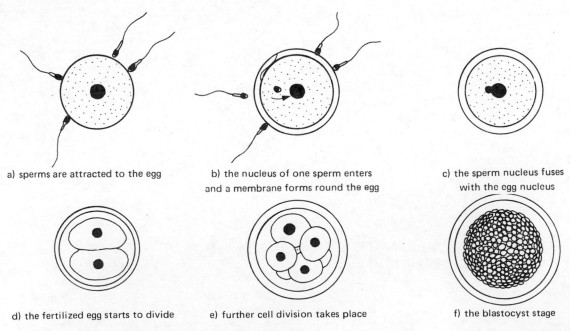

a) sperms are attracted to the egg

b) the nucleus of one sperm enters and a membrane forms round the egg

c) the sperm nucleus fuses with the egg nucleus

d) the fertilized egg starts to divide

e) further cell division takes place

f) the blastocyst stage

Fig. 3:12 Fertilization and early development of the egg.

takes place, that is, the follicle bursts and liberates the egg into the **fallopian funnel** at the end of the fallopian tube. The egg is wafted along by the action of the ciliated cells which line the funnel and tube. The egg is still surrounded by a layer of cells derived from the follicle and is of a size just visible to the naked eye. It may not survive much longer than a day unless it is fertilized. Meanwhile, cells of the follicle from which the egg has been expelled divide rapidly to form a gland composed of yellowish cells, the **corpus luteum** (p. 35).

If it is fertilized the egg (now a zygote) starts dividing into 2, 4, 8 cells etc. until it becomes a hollow sphere of cells called a **blastocyst** (Fig. 3:12f). By this time it will have completed its passage down the tube and reached the uterus. Occasionally, during the first division of the zygote, the two cells which are formed separate completely and continue development quite independently; in this way **identical twins** are formed. They are called identical because they come from the same zygote and therefore have exactly the same hereditary characteristics.

Implantation and development

The blastocyst, on reaching the uterus, sinks into its spongy, vascular walls and becomes surrounded by maternal tissues, a process called **implantation**. From then onwards the developing embryo is dependent upon the blood of the mother for nutrients and oxygen.

Cell division continues rapidly and while some cells develop into the tissues and organs of the developing baby, now called a **foetus**, others form membranes: these include the **amnion** and those which help in the formation of the **placenta**. The amnion is a protective membrane which projects into the cavity of the uterus and encloses the foetus in a bath of fluid. The latter acts as a shock absorber, protecting the foetus from mechanical injury and allowing it some freedom of movement when it becomes larger.

The placenta, when fully developed, is a complex organ formed partly from the wall of the uterus and partly from tissues derived from the embryo. It forms a large disc-shaped structure richly supplied with blood vessels. The foetus is connected to the placenta by the **umbilical cord**. After about a month's de-

velopment the foetus has its own blood circulation; this includes an artery which passes up the umbilical cord, breaks up into a mass of capillaries in the placenta and returns as a vein to the foetus. In the placenta the capillaries of the foetus become very closely associated with the blood vessels of the mother, but are never actually connected with them (Fig. 3:13). Even so, oxygen and soluble food material can easily and quickly pass from mother to foetus and carbon dioxide and nitrogenous waste can pass from foetus to mother. In this way the foetus is able to obtain all that it requires for growth until it is born.

The substances that the foetus receives are selected to some extent by the placenta; most

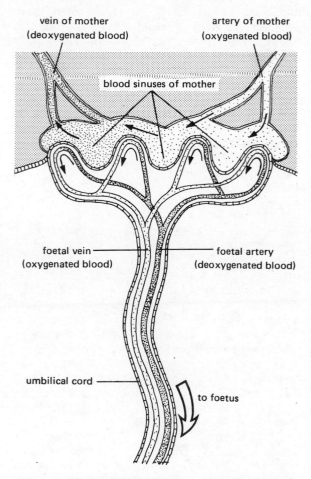

Fig. 3:13 Much simplified diagram showing how the circulations of mother and foetus associate in the placenta.

31

substances with large molecules are unable to pass through, so the foetus is protected from those in the mother's blood which might otherwise harm it. However, this protection is not perfect. Nicotine from tobacco causes rapid stimulation of the heart of the baby and in those mothers who smoke a lot during pregnancy, nicotine may adversely affect the baby's growth. Alcohol and other drugs also have their hazards. For example, in the early 1960's the taking of the drug thalidomide as a sedative during pregnancy tragically resulted in a large number of babies being born with deformities.

As pregnancy proceeds the uterus enlarges considerably to accommodate the developing foetus and its amniotic sac (Fig. 3:14), causing some of the abdominal organs to be displaced. The walls of the uterus also become much more muscular in readiness for birth when their contractions help to expel the baby from the body. The breasts also develop, the glands within them enlarging in preparation for the secretion of milk. The foetus is capable of a certain amount of movement. At about $4\frac{1}{2}$ months the mother can feel her baby kicking, and at a later stage the baby may even suck its thumbs or scratch its face with its growing finger nails. Shortly before birth the position of the baby alters so that it lies head-downwards near the opening of the cervix (Fig. 3:15).

Birth

Hormones secreted by the pituitary gland influence the onset of birth or **labour**, when the walls of the uterus start to contract rhythmically. The contractions are slight at first and intermittent, but as labour proceeds the contractions increase in strength and frequency. Sometime during labour the amnion breaks and the clear amniotic fluid escapes through the vagina. As labour proceeds the opening of the cervix gradually dilates as the baby's head presses against it, and the contractions of the mother's abdominal muscles assist those of the uterus to help force the baby towards the outside world.

At birth the baby takes its first breath, the lungs expanding from their shrunken condition; this produces its first cry. The baby is still attached to the placenta by the umbilical cord; this is tied to prevent bleeding and then cut.

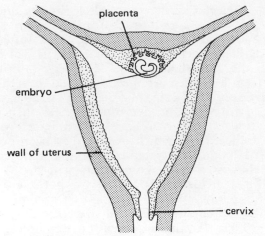

a. SOON AFTER IMPLANTATION

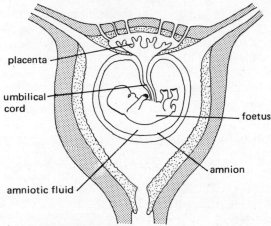

b. AT 6 WEEKS

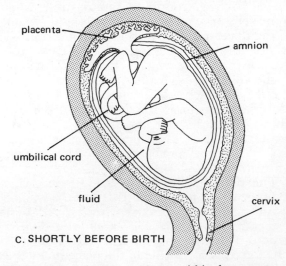

c. SHORTLY BEFORE BIRTH

Fig. 3:14 Development of foetus within the uterus.

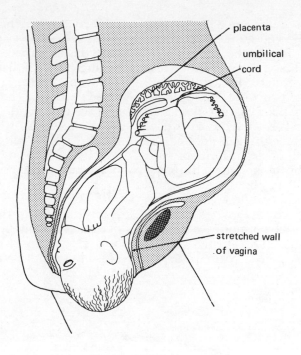

placenta

umbilical cord

stretched wall of vagina

Fig. 3:15 Birth.

Soon after birth the placenta becomes detached from the wall of the uterus and is expelled as the **afterbirth**. The navel, which everyone has in the centre of the abdomen, marks the place where the umbilical cord was attached.

The whole period from fertilization to birth is called the **gestation** period; it is nine months in humans. After the first two months, when the main body systems are laid down, the rest of the time is largely spent in growth. At birth the average weight is 3·2kg (about 7lb). Occasionally a pregnancy does not persist and the uterus evicts the foetus during early development; this is called a **miscarriage** or **abortion**. Some babies may occasionally be born prematurely for some reason, but their survival depends largely on their size; if they are over 2kg they have a fair chance of survival, given specialised medical care.

For the baby, birth involves great changes. Previously it was dependent on its mother's circulation for food and oxygen, now it has to use its alimentary canal and lungs for the first time; previously its temperature was controlled by the mother, now it has to use its own heat-regulating system; this takes a little while to adjust.

Lactation

The baby usually takes little nourishment for the first day or so although it has a natural instinct to suck. This sucking stimulates the mother's breasts to produce milk. The taking of the milk into the stomach is a new experience, but gradually the digestive glands adjust and secrete their juices efficiently and the baby becomes used to its new method of feeding. For the first days the breasts secrete a clear fluid called **colostrum** which is more easily digested than milk. It is also rich in antibodies and is of great value to the baby.

The influence of hormones on reproduction

The ovaries and testes not only produce gametes, but they also act as ductless glands and secrete hormones. The testis secretes the male hormone, **testosterone,** and the ovary the female hormone, **oestrogen**. These are called **sex hormones** as their main function is to bring about the development of the **secondary sexual characters**, that is, the characteristic features which distinguish males from females.

33

The sex of a baby is determined at the time of fertilization (p. 171) and quite early in the development of the foetus the appropriate gonads are formed. These secrete either testosterone or oestrogen, according to the sex, and these hormones influence the development of either the male or female characters so that when the baby is born it has the basic features of either a boy or a girl. During childhood the gonads remain more or less dormant and secrete very little sex hormone. But at the onset of puberty, as a result of stimulation from a hormone from the anterior pituitary, the gonads not only enlarge and produce gametes but they start secreting their sex hormones in much greater concentrations. This brings about further changes associated with puberty.

Puberty starts in the male in the early teens, and in the female a year or so earlier.

In the male there is a rapid increase in height during puberty, hair grows on the face, chest and pubic region, the larynx enlarges and the voice deepens. The reproductive system itself also develops and becomes functional. In the female there is also rapid increase in height, the breasts develop, the pelvic (hip) girdle widens, the reproductive system becomes functional and **menstruation** starts.

The ovarian and menstrual cycles

Reproduction in the female is restricted to a period from puberty to **menopause** (about 50). The latter is the time when the ovaries cease to produce any more eggs.

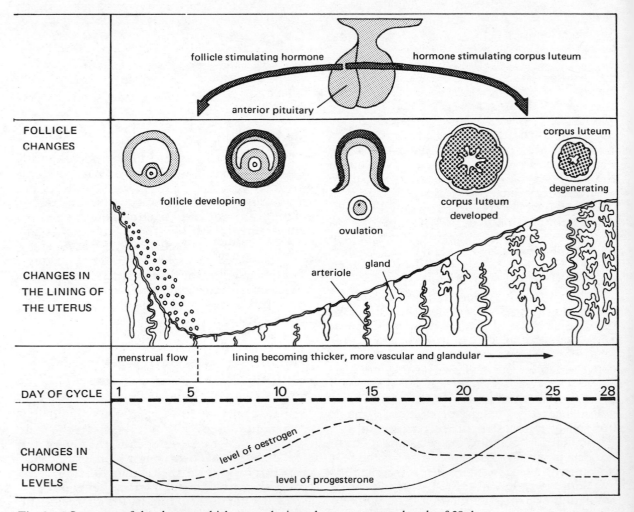

Fig. 3:16 Summary of the changes which occur during a human menstrual cycle of 28 days.

The **ovarian cycle** usually lasts 28 days and can roughly be divided into two halves. During the first 14 days, stimulation of the ovary by a pituitary hormone causes a Graafian follicle to become mature and an egg is shed. In the same period the ovary is stimulated to produce its own hormone, oestrogen. During the second fortnight, cells in the follicle from which the egg was shed quickly multiply to form a gland, the **corpus luteum**. This gland is stimulated by another hormone from the pituitary to secrete a further hormone, **progesterone**. So during the monthly cycle the levels of oestrogen and progesterone in the blood fluctuate. During the first fortnight there is a build-up of oestrogen, and in the second, an increase in progesterone (Fig. 3:16). This fluctuation of oestrogen and progesterone causes changes to occur in the lining of the uterus; these changes constitute the **menstrual cycle**.

By the time ovulation takes place the lining of the uterus is spongy and vascular and ready to receive an egg if it is fertilized. If the egg is *not* fertilized, the lining cells break away, the egg is passed out of the body and there is some bleeding. This **menstrual flow** or **period** lasts four to five days. If, however, the egg *is* fertilized, the corpus luteum persists and continues to secrete progesterone which exerts a further influence on the growth of the uterus and the breasts. It also has the effect of suppressing further ovulation and causing the cessation of menstrual periods—usually the first sign that pregnancy has begun.

Social implications of reproduction

When a mammal becomes mature the hormones circulating in its blood also affect its behaviour. This is often shown as complex patterns of courtship, mating and preparation for the birth of the young. This also happens in humans, but there are important differences. Some mammals such as foxes and deer breed only once in a year, others such as rabbits and voles have several litters, but man has no breeding season, and can reproduce at any time between puberty and menopause. Consequently we have to come to terms with our sex drive because we become physically mature before we are fully developed emotionally and intellectually. Bearing children is a considerable personal and social responsibility for both mother and father.

Being responsible for a new life involves the provision of suitable conditions for the child's development and humans have an extremely long period of dependence on their parents.

There is overwhelming evidence that marriage and a stable home background provide the best conditions for a child's development. In this way the child receives the love and security needed for full emotional development and can use to the full the long period of learning while he or she is dependent. Within the family the child also learns the art of giving as well as receiving, which is the pre-requisite of a happy life, and for taking social responsibility in the larger society later on.

Child-bearing also has social implications in another way. A large family can be a happy and stable community, but frequently the children are deprived of the care, understanding and financial resources that they need, and many problems arise for both the children and their parents. Furthermore, with a marked population explosion in the world, it is especially important to make sure that every pregnancy is a desired one.

Family planning

Responsible parenthood today demands that the family is properly planned. A happy, secure family is more likely to be created if the children are wanted in the first place. For this reason many people advocate some form of birth control (contraception) to prevent unwanted pregnancies, although others disapprove of this practice on religious grounds.

Various contraceptive methods are practised, some of which aim to prevent sperms from reaching an egg during sexual intercourse; another one known as the 'pill' affects the ovarian cycle. It contains a mixture of two hormones, progesterone and oestrogen in the correct proportion to prevent ovulation (see Fig. 3:16). When the pill is no longer required the normal cycle is again possible. These hormone pills, of which there are several kinds, are only taken under medical supervision.

Artificial abortion

This involves the removal of the foetus and the consequent death of the potential baby. It is looked upon by far too many as an easy way out of an unwanted pregnancy, but in fact it is an operation of considerable danger unless performed by an experienced surgeon. An abortion may also lead to sterility and, in addition, it often has serious psychological repercussions, because child-bearing is such a basic instinct in a woman.

Opinions are greatly divided about the ethics of abortion, but all would agree that prevention of a pregnancy is better than destroying a developing child. The number of legal abortions carried out in Britain in 1971 exceeded 182,000. Illegal abortions might even increase that figure by 50,000. This reflects the price that is being paid for irresponsibility in sexual relationships.

Diseases related to the reproductive system

These are termed **venereal** diseases (VD). There are two of considerable importance, **gonorrhoea** and **syphilis**, both of which are caused by a bacterium and are spread by an infected person through sexual intercourse. Recent medical statistics show that the number of people contracting these diseases is increasing each year at an alarming rate and bringing untold misery. For example, there were 55,000 new cases of gonorrhoea in Britain alone during 1971, and of these, 10,000 occurred amongst teenagers. This appears to be due to the great increase in promiscuity (sleeping around).

With gonorrhoea, some women do not, at first, show any obvious symptoms, so if promiscuous they can spread the infection. In the later stages of the disease the infection spreads through the reproductive system causing severe illness and sterility. A pregnant woman with the disease may also infect her child during labour and its sight may be endangered in consequence. In men the infection starts in the penis and may spread to the bladder and the kidneys with serious results.

With syphilis all parts of the body can be affected as the micro-organisms pass into the blood stream; infection of the foetus can also occur.

Both diseases have been treated effectively with antibiotics and when penicillin was discovered the incidence of these diseases dropped; however, today some strains of the organisms causing gonorrhoea have become resistant to penicillin and this has raised serious problems.

4

Reproduction in flowering plants

Flowering plants use both asexual and sexual methods of reproduction. The asexual method is also known as vegetative reproduction as it takes place by means of buds; sexual reproduction is brought about through the formation of flowers and the production of seeds.

ASEXUAL OR VEGETATIVE REPRODUCTION

Flowering plants reproduce vegetatively in many ways, but the principle involved is always the same; somewhere on the parent plant a bud develops and eventually becomes separated to form a new plant. At first the bud is always dependent upon food obtained from the parent, but later it develops foliage leaves of its own and so is able to photosynthesise for itself. Adventitious roots (i.e. those formed from a stem) also develop and these enable the young plant to absorb water and salts. Thus when separation from the parent occurs the new plant is completely self supporting. Different species form these reproductive buds from different organs. Some of these organs can be summarised as follows:

Modified stems (Fig. 4:1)

Rhizomes
These are stems which grow horizontally below ground level. They produce buds, some of which may grow out into side branches and come above the ground as shoots. Eventually the newly-formed rhizomes become separated from the parent by rotting and reproduction is finally effected. Examples include couch-grass, Michaelmas daisy, mint and iris.

Tubers
These are the swollen ends of underground stems. Potatoes and Jerusalem artichokes can reproduce from these.

Runners
These are stems which grow along the surface of the ground and root at intervals. In the strawberry they arise as axillary shoots and their terminal buds grow out to form new plants; adventitious roots develop at the same point. Other runners can arise similarly from the daughter plants. Eventually the connecting stems rot away and the daughter plants become independent.

Corms
These are short, erect underground stems. Crocus, gladiolus and montbretia are examples. Superficially they look like bulbs, but a true bulb contains fleshy leaves and is not solid (Fig. 4:3).

Modified roots

Root tubers
These are swollen adventitious roots and are found in lesser celandine, dahlia and many orchids. New plants may grow from buds which arise near their point of origin with the stem (Fig. 4:1).

Modified buds

Bulbs
These are buds which have much stored food in some of their scale leaves. When they reproduce, new bulbs arise from buds in the axils of these scale leaves. Daffodils, bluebells and onions are familiar examples (Fig. 4:1).

Modified leaves

Some plants, such as *Bryophyllum*, form buds from the edge of the leaves which quickly grow into tiny plants complete with adventitious roots. They drop off, the roots grow into the soil and new plants develop (Fig. 4:2).

Examine a selection of these organs of vegetative reproduction in more detail and grow them to see how they reproduce.

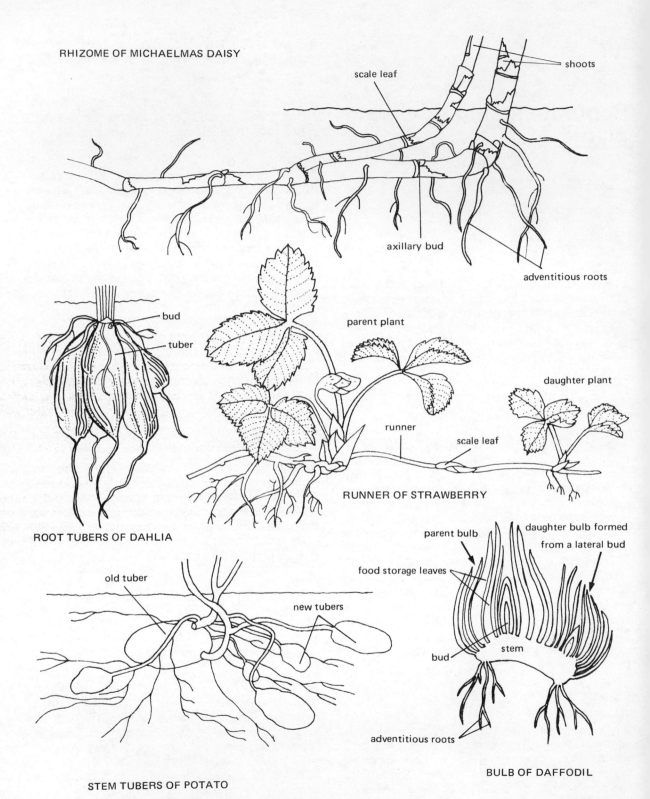

RHIZOME OF MICHAELMAS DAISY

scale leaf

shoots

axillary bud

adventitious roots

bud

tuber

parent plant

daughter plant

runner

scale leaf

RUNNER OF STRAWBERRY

ROOT TUBERS OF DAHLIA

parent bulb

daughter bulb formed
from a lateral bud

food storage leaves

old tuber

new tubers

bud

stem

adventitious roots

STEM TUBERS OF POTATO

BULB OF DAFFODIL

Fig. 4:1 Some organs of vegetative reproduction in flowering plants.

Potato—a stem tuber. Examine a potato and note:

1. The 'eyes' which are the buds lying in the axils of curved scale leaves (or their scars).
2. The concave side of the scale leaves all point towards one end where the growing point is situated.
3. At the opposite end to the growing point is a scar where the tuber was attached to the parent plant.
4. The tough skin which is made of cork.
5. Many brown dots scattered over the skin. These are **lenticels**, small pores in the skin through which respiratory gases can diffuse.

Grow a small potato in earth against the glass side of a container so that you can see what happens when it grows, *or* grow one in

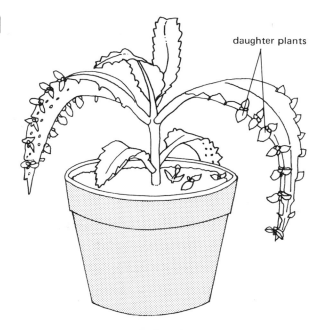

Fig. 4:2 (right) *Bryophyllum*: miniature plants develop at the margins of the leaves and later drop off and grow.

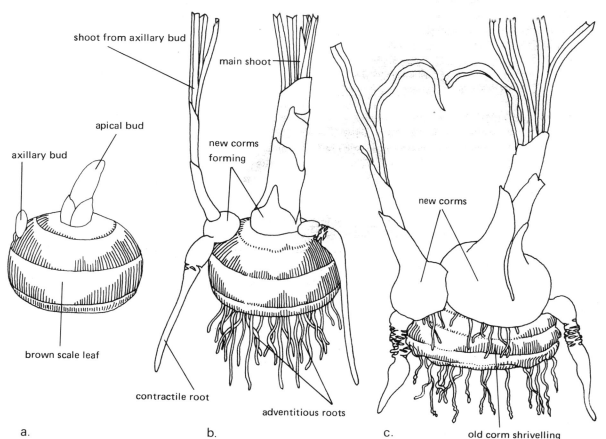

Fig. 4:3 Stages in vegetative reproduction of a crocus corm: a) winter b) early spring c) late spring.

39

light soil and after some weeks, when the foliage has grown well, dig it up very carefully so as not to brush off the new potatoes that have formed, and examine the whole plant.

Which buds have grown into shoots? From which parts do the adventitious roots grow? Where do the stems which bear the new potatoes arise? Can you think of a good reason why gardeners bank up earth round the base of the green shoots when they have grown fairly large?

Crocus—a corm. Examine a crocus corm and note:
1. The fibrous brown scale leaves which cover it.
2. The scars that are left when you pull off the scale leaves. Can you find any axillary buds just above these scars?
3. One or more larger buds at the top of the corm. Slice one down vertically and find the tiny leaves and flower bud encased in protective scale leaves.
4. Cut the corm itself and note how it is composed of solid stem. Add a drop of iodine to the cut surface. What food substance does it contain?

Plant another corm in a small flower pot containing soil so that the corm is only half covered; water it regularly and study its development.
1. From which parts do the adventitious roots grow? What is their function?
2. When the shoots grow, observe how the leaves and flowers develop. Do any of the axillary buds grow out from the side?
3. After flowering, do any swellings arise at the base of the shoots? If so, these will be the beginnings of new corms. Where does the food come from that makes them swell?
4. Where do the scale leaves which will surround these new corms come from?

Large adventitious roots will also grow out from the new corm; they are contractile and will eventually pull the new corm further into the ground.

Artificial methods of vegetative reproduction

We can make use of these methods of vegetative reproduction in a number of ways:

Separation or division
In the autumn, plants such as dahlias are dug up and their tubers are separated to increase the number of plants for the following year. All plants having rhizomes, for example Michaelmas daisy, can be treated in a similar way to form new clumps. Unfortunately, when the soil is dug the rhizomes of certain weeds such as bindweed and couch-grass may get cut up into many pieces all of which could grow into new plants; hence the need for pulling them out without breaking them.

Layering
Some plants are stimulated to put out adventitious roots when a shoot comes in contact with moist earth. By pegging down the shoots of plants such as carnations this reaction will quickly take place and the new plants can later be separated (Fig. 4:4b).

Cuttings
The aerial shoots of many plants can also be cut off and planted in damp sandy soil to stimulate the formation of adventitious roots. Figure 4:4a shows the technique for doing this. Geraniums, chrysanthemums, fuchsias and many other plants are normally propagated in this way. Some plants which will not readily form roots on their cuttings may be helped to do so by treating the cut ends with a hormone powder; this stimulates their growth (p. 69).

Budding
This is the standard method of propagating roses. The principle is to take a bud from the rose to be propagated, called the **scion**, and insert it beneath the bark of the stem of a wild rose plant, called the **stock**. To be successful, the cambium tissue of each must be against that of the other. Why the cambium?

Budding should be done in the autumn after the scion has flowered. Using a sharp knife a leaf and its dormant axillary bud is cut out together with a thin slice of wood behind it (Fig. 4:5a). This wood is carefully removed to expose the cambium underlying the bud. A T-shaped cut is then made in the bark of the wild rose stock and the flaps peeled back to allow the scion to be inserted below it; it is then lightly bound in place. Finally the leaf is cut off leaving the bud and a short piece of petiole.

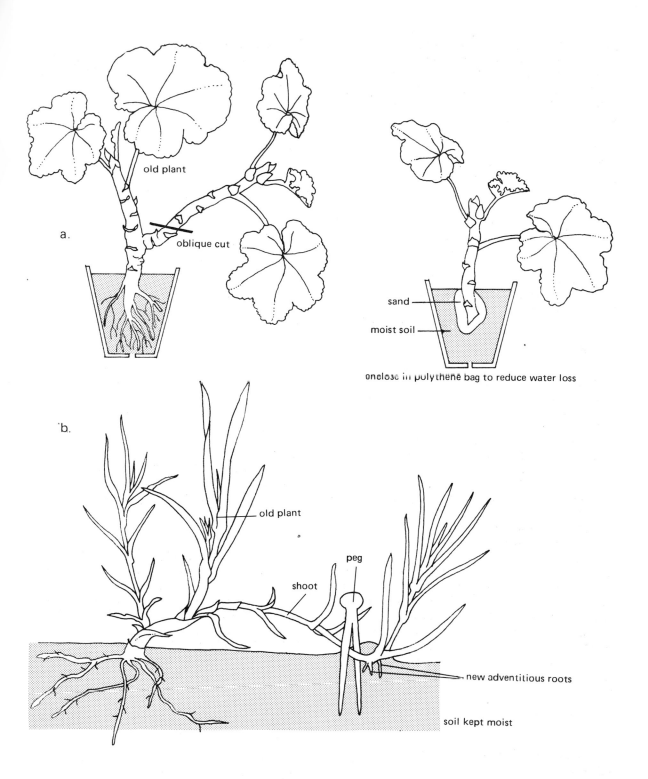

old plant

a.

oblique cut

sand

moist soil

enclose in polythene bag to reduce water loss

b.

old plant

peg

shoot

new adventitious roots

soil kept moist

Fig. 4:4 Artificial methods of vegetative propagation: a) by taking cuttings (geranium) b) by layering (carnation).

41

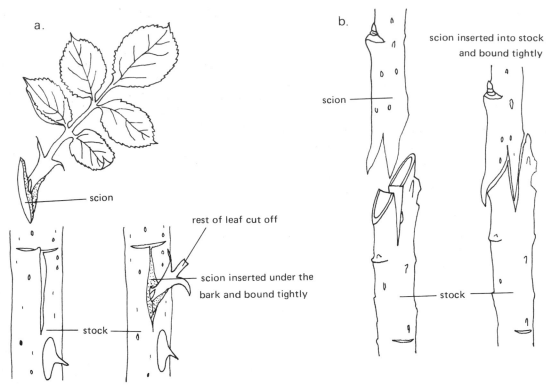

Fig. 4:5 a) Method of propagating a rose by budding. b) Method of grafting an apple twig.

In the spring it should grow rapidly and the stock shoot above the bud can be cut off.

Grafting

This method is much used for propagating fruit trees. In principle it is rather similar to budding but the scion used is a woody twig which is cut in such a way that its exposed cambium comes in close contact with the cambium of the stock on which it is grafted. The two are bound firmly together (Fig. 4:5b). Buds on the stock below the graft should be removed. Why do you think this should be done?

Perennation and food storage

Perennation is the ability to survive the difficult conditions of winter. Those plants which are able to do this are called **perennials**. The two requisites for survival are a means of protecting the delicate growing regions from cold, dry winds and frost, and the provision of enough food for growth in the spring before the new leaves take over this function.

Herbaceous perennials are those which have shoots which die down each winter, but also have special underground portions which survive. The buds on these are thus protected from severe frost. In the spring they form new aerial shoots. Rhizomes, tubers, bulbs and corms act in this way and are therefore perennating organs as well as organs of vegetative reproduction. You will have noticed that they also act as food stores; this enables growth to occur in the spring before the foliage leaves have been formed. This food store makes early flowering possible, a great advantage in the case of woodland plants as they can both flower and photosynthesise before the trees and shrubs above them come into leaf and cut off much of the light. Why do you think it is advantageous for flowering to occur during the well-lit period?

Woody perennials include trees and shrubs. These have permanent woody portions above ground. Many are **deciduous**, shedding their leaves in the autumn; others are **evergreen**, their leaves persisting through the winter as they have adaptations for reducing the rate of transpiration (p. 18). Woody perennials in cold climates protect their growing points by having winter buds, their scale leaves giving special protection from frost.

Some plants are unable to cope with winter conditions and survive only as seeds. These are the **annuals** which complete their life cycle within the year. **Biennials** take two years to complete their life cycle. During the first year they germinate and produce much leafy growth which forms the food; the latter is then stored in underground roots, usually causing them to swell up in consequence. During the winter the aerial shoots die, but the following spring the stored food is used in new growth with the formation of flowers, fruit and seeds before the plant dies. Carrots, parsnips and turnips are examples of biennials (Fig. 4:6).

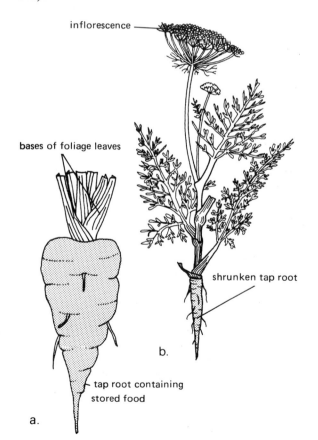

inflorescence

bases of foliage leaves

shrunken tap root

b.

tap root containing stored food

a.

Fig. 4:6 Carrot—a biennial: a) at the end of the first year × ½ b) forming flowers and fruits during the second year × ⅙.

SEXUAL REPRODUCTION IN FLOWERING PLANTS

The flower is the part of the plant which is specially adapted to bring about sexual reproduction (Fig. 4:7). Flowers vary greatly in size and structure, but basically they all have a common plan. This consists of several **whorls** (rings) of modified leaves arising from the swollen end of the flower stalk called the **receptacle**. In most flowers four whorls are present:

1. The calyx
This is the outer whorl and is composed of **sepals** which protect the rest of the flower when in bud.

2. The corolla
This lies within the calyx and is composed of **petals**. In flowers which are insect-pollinated they are usually coloured to attract insects and may act as a platform for them to settle on.

3. The androecium
This consists of one or more whorls of **stamens** which produce the **pollen grains** in which the male gametes develop. Each stamen consists of a stalk or **filament** and two **anthers** at the top which produce the pollen.

4. The gynoecium
This is the centre of the flower and is made up of one or more **carpels**. The latter form and protect the **ovules** which contain the female gametes. Each carpel consists of a basal part, the **ovary**, which contains the ovules, a special portion at the other end, the **stigma**, which receives the pollen grains during pollination, and a connecting part, the **style**. The ovules after fertilization develop into **seeds** and the carpels become **fruits**.

In insect pollinated flowers, structures called **nectaries** may be present; these secrete a sugar solution called **nectar** to which insects are attracted. Nectaries often occur at the base of the carpels but in some flowers they may be modifications of the petals.

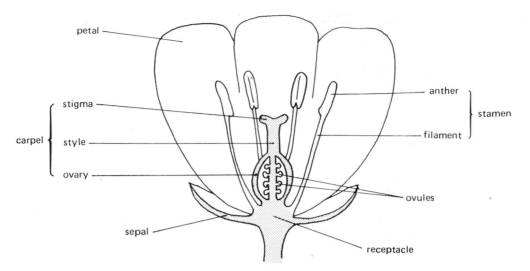

petal

stigma

carpel

style

ovary

sepal

anther

stamen

filament

ovules

receptacle

Fig. 4:7 The main parts of a flower.

Variation in flower structure

In the course of evolution the basic plan of flower structure has become greatly modified, and if you consult an identification book you will see how flowering plants are classified into many families. When examining flowers you should notice the number of the parts; they are often in multiples of threes, fours or fives. The flowers may be regular, or irregular; some form open cups, others are tubular; some have their parts fused together, in others they are separate. Examine a selection to see how they vary; we will just look at three in detail.

Wallflower

This is a member of the family *Cruciferae* to which cabbages and cresses also belong (Fig. 4:8).

First examine the whole **inflorescence**, i.e. the whole group of flowers on a stem. Note that the buds or younger flowers are at the top, then come the flowers which are fully out and below these are the flowers which are withering or are completely over. The latter may have produced long fruits.

Now examine a single flower. Note the calyx on the outside; it is composed of four sepals in opposite pairs. (Look also at a bud and see how the sepals enclose and protect the other parts of the flower.)

Remove the sepals with forceps to expose the corolla completely. Each petal has a flat part above which is coloured and acts as a platform on which insects can land, and a large claw which attaches the petal to the receptacle. The four petals are also arranged in opposite pairs. Do they alternate with, or come opposite to, the sepals? Remove the petals to expose the androecium. Notice the position and relative lengths of the stamens—the two outer ones are shorter than the four in the middle.

Examine a stamen under a lens. Note the long filament and the two anthers at the end lying side by side and separated by a groove. Remove a stamen and examine its anthers under a microscope. Look for the pollen grains adhering to them.

Now remove the remaining stamens to expose the gynoecium in the centre. Look for green, glistening bumps on the receptacle near the base of the gynoecium; these are the nectaries which secrete nectar on which bees and butterflies feed. How many are there?

Examine the gynoecium. The long cylindrical portion is the ovary; it tapers slightly to form a short style and ends in two stigmas. Cut the ovary transversely and look at the cut ends; it is divided into two compartments. Cut an ovary from another flower lengthwise between the stigma lobes and note the row of ovules which will develop into seeds if fertilization occurs. The structure of the gynoecium can be explained by the fact that it is composed of two carpels fused together, but the stigmas have remained separate.

44

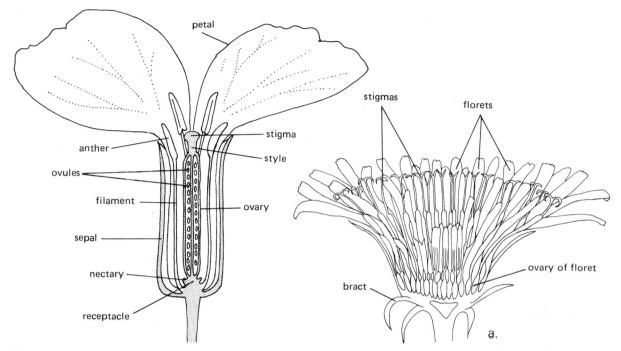

Fig. 4:8 Wallflower: half flower.

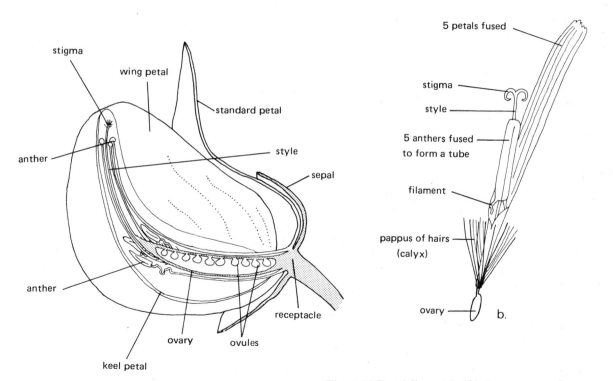

Fig. 4:9 Lupin: half flower.

Fig. 4:10 Dandelion: a) half inflorescence × 2
b) single floret × 4.

Lupin

This is a member of the family *Papilionaceae* to which peas, beans and laburnum also belong (Fig. 4:9).

Note that in this inflorescence, like that of the wallflower, the buds are at the top and the flowers which open first are the lowest ones.

Examine a single flower and note how irregular it is compared with that of the wallflower. Instead of being separate, the sepals are all fused together to form two lips and the petals are of varying shapes. The large petal which stands up at the back is called the **standard**; it makes the flower conspicuous and helps to attract insects. The two at the sides are the **wing** petals which act as a platform for insects, and in the centre are the two **keel** petals fused together to form a tube with a small hole at the end. Remove the petals carefully and note how wing and keel petals fit together as in a ball and socket joint. Inside the keel petals you will find both androecium and gynoecium. The former consists of ten stamens which are united by their filaments to form a sheath. Carefully cut this away and examine it. The gynoecium is now exposed in the middle; it consists of a single carpel, the ovary being pod-like and giving rise to the long style with the stigma at the end. Cut the ovary longitudinally. Note the row of ovules.

Dandelion

This is a member of the *Compositae* family along with hawkweeds, daisies and thistles (Fig. 4:10).

Examine a dandelion and see if you can find the sepals, petals, stamens and carpels. You will find this a puzzle. The clue to the problem is that you are looking at a whole inflorescence and not just one flower. The green structures at the base are *not* sepals, but modified leaves called **bracts**, and the yellow structures are *not* single petals but complete flowers, usually called **florets**.

Cut the inflorescence longitudinally and you will be able to separate the individual florets. Examine one under the low power of a microscope. There are no sepals as such, but the calyx is reduced to a ring or **pappus** of hairs. The corolla is composed of petals all fused together. You can find out the number by counting the teeth at the end. Note how the corolla is tubular at the base but flattens out higher up to become strap-shaped. The five stamens have their anthers fused together to form a cylinder round the style. Below the anthers you should be able to see the five filaments which are attached to the corolla. The gynoecium consists of a single ovary at the base of the floret from which a long style projects; it passes through the stamen sheath and divides into two stigmas. The ovary contains a single ovule and because the rest of the flower parts arise from a position *above* it, the ovary is said to be **inferior**. This is in contrast to the wallflower and lupin where the parts arise from *below* the base of the ovary; in these the ovary is said to be **superior**.

If you examine a daisy you will find that, in contrast to the dandelion, there are two kinds of florets, strap-shaped ones on the outside called **ray florets** and short tubular ones in the centre known as **disc florets**. The ray florets have no stamens.

Pollination

Flowers have to be pollinated before their ovules can be fertilized and seeds produced. Pollination is the transference of pollen from the stamens of one flower to the stigmas of either the same flower—**self-pollination**—or more often another flower of the same species—**cross-pollination**.

Fertilization occurs later when male and female gametes fuse together. To make fertilization possible, the pollen which produces the male gametes and the ovules which form the female gametes must be brought near enough to each other for them to meet. That is why the transfer of pollen from stamen to stigma must take place first. Later (p. 51) you will see how the gametes are actually brought together for fertilization.

Pollination may be effected by many agents. For example, if we walk through a buttercup field our shoes become covered in pollen and no doubt some is rubbed off on to the stigmas of the flowers. The chief pollinators are insects and wind.

Insect pollination

To make this effective:
1. Insects must first be attracted to the flower in some way. This requirement is met by

flowers which provide food in the form of nectar or abundant pollen.

2. The flower must be easily found—it pays to advertise! For this reason insect-pollinated flowers are often large and brightly coloured which make them conspicuous; they also have a characteristic scent. (The senses of sight and smell in insects are very highly developed.)

3. Pollen must not be wasted. By having pollen grains with rough coats, the grains tend to clump together and therefore cling better to the hairy legs and bodies of insects; this ensures that not too much pollen is lost. Also, by having a scent which is characteristic of the species, the flower will be recognised more easily by an insect. As many insects are attracted to flowers with a particular scent, it is more likely that the pollen will reach a flower of the same species.

4. When the pollen grains reach the flower, they must have the best chance of actually being deposited on the stigma—a small target. Hence the position of both stamens and stigma is important in relation to the position of the alighting insect.

Let us see how these conditions are fulfilled in the flowers whose structure we have already studied.

Wallflower

The flowers are brightly coloured and highly scented. The nectar at the base of the stamens can be reached mainly by insects with a long proboscis such as butterflies and bees. The stamens and stigmas are arranged at the top of the tube formed by the four petals. When the insect lands on the petals and puts its proboscis down the tube, its head rubs against the stamens and some pollen is brushed on to it. If it then visits an older flower, the style will have grown longer and the stigmas will project *beyond* the stamens, the best position to receive the pollen (Fig. 4:11).

Lupin

The inflorescence is very conspicuous and, although no nectar is produced, a great deal of pollen is formed. The flowers are visited mainly by bees. In a young flower the stamens inside the keel produce a lot of pollen which collects at the end of the keel. When the bee

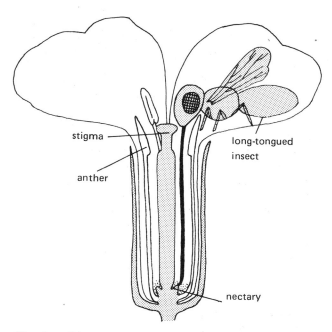

Fig. 4:11 Diagram showing how pollen is effected in a wallflower.

alights on the wing petals its weight depresses them, and because of the ball and socket joint between them, the keel petals are also pulled down causing the stamens and stigma to act like a piston forcing the pollen out of the hole at the end of the keel on to the underside of the bee's body. (Examine a young flower yourself and pull the keel down and see how this happens.) As the flowers become older, the style gets longer, so that when a bee alights on an older flower it is the stigma that hits the underside of the bee and so receives pollen. Self-pollination is avoided because pollen will not develop on the stigma of the same flower (Fig. 4:12).

Dandelion

Individual florets are small, but the whole inflorescence is very conspicuous, highly scented and produces much nectar. The florets are pollinated mainly by bees and butterflies. The pollination mechanism is adapted for cross-pollination but self-pollination can occur if this fails.

Take a dandelion which still has florets in bud in the centre. Select three florets as follows: one from the centre, one from further out

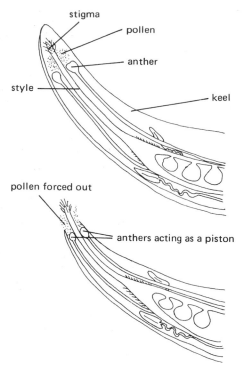

Fig. 4:12 Pollination in the lupin: (top) central part of flower cut open (bottom) the pollen is pushed out of the end of the keel when the bee alights.

Fig. 4:14 Bee pollinating a dandelion.

Fig. 4:13 Pollination mechanism in the dandelion.

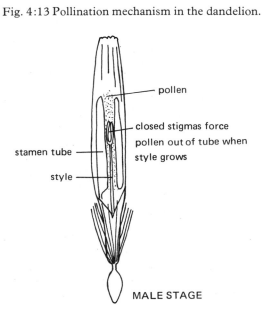

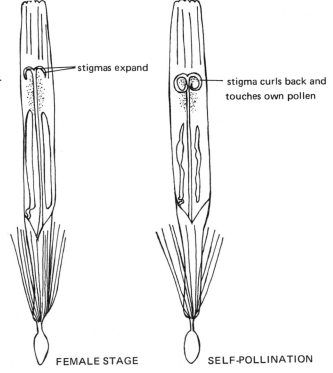

48

and an old one from the outside. Examine all three under the low power of the microscope and note the differences in the position of the stigmas and in the areas where the pollen grains are clustered.

The explanation of what you have seen is that in young florets the stamens shed their pollen into the stamen sheath, and as the style grows it acts like a piston and forces the pollen out of the top. At this stage the two stigmas are pressed closely together and cannot be self-pollinated. Later the stigmas expand and can receive pollen from another floret if an insect comes in contact with them. If this is not effected, then, later still, the stigmas curl round and touch the hairy style to which the floret's own pollen adheres and self-pollination may occur (Fig. 4:13).

This excellent method is used by all members of the *Compositae* but in the dandelion, in spite of the elaborate mechanisms, pollination has in the course of evolution become unnecessary as the seeds develop without any fertilization at all, a process called **parthenogenesis**. This has been proved by taking a dandelion before the flowers have opened and cutting off all the parts of the florets above the ovaries with a razor. In spite of this treatment the ovules still developed into seeds capable of germination.

On a sunny day watch how other flowers are pollinated. Note the types of insects which visit them, and whether they are long- or short-tongued. How is the shape of the flower adapted to receive these particular insects? Good flowers to watch are those of fruit trees, antirrhinums, dead-nettles and poppies.

Can you explain the following?
1. a) Certain flowers give out their scent mainly in the evening; b) flowers which do this are usually white.
2. Bees visit certain flowers which appear red to us, yet they are unable to detect red as a colour.
3. Some plants may not form fruits well when they are grown in greenhouses. (Can you also think of a remedy for this?)

Wind pollination

Flowers which are wind-pollinated are very different in structure. They include grasses, nettles, plantains, shrubs such as hazel and many trees including oak and ash. Obviously, the wind does not have to be attracted to the flowers, so they are often inconspicuous, lack scent and produce no nectar. As wind may blow the pollen in any direction there will inevitably be great wastage, and because of this great quantities of small, light pollen grains are produced in order that some may reach their target. For example, when in flower, grasses produce so much pollen that people who are allergic to pollen may develop hay fever at that time, even if they live in towns. The likelihood of the pollen reaching its target in wind-pollinated flowers is greater if the species is a common one which normally grows over wide areas or in clumps.

The flowers obtain greater advantage from the wind if they are placed in a position where they are easily blown. A common arrangement is for the flowers to be grouped in hanging catkins, e.g. hazel. In catkins, the stamens hang loosely outside the flowers, allowing the wind to blow them. For the same reason, the stigmas of wind pollinated flowers, by being large, feathery, and in an exposed position have more chance of receiving the blown pollen. Why do you think wind pollinated flowers often come out before the leaves appear?

Examine the flowers of one of the grasses and see to what extent the above characteristics are true for that species. Grass flowers are extremely small and grouped together in very compact inflorescences. Examine the inflorescence first and note how the stamens and stigmas project, and how large the anthers are. Now dissect out a single flower with a needle and look at it under the low power of the microscope. There are no sepals or petals to be seen and the stamens and gynoecium are partly protected by two green bracts (Fig. 4:17).

Avoiding self-pollination

The majority of flowers have some device which makes self-pollination less likely or impossible. Here are some examples:
1. The stamens ripen before the stigmas, e.g. buttercup.

49

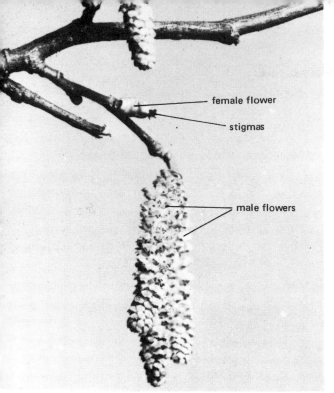

Fig. 4:15 Hazel: a wind pollinated shrub with catkins.

Fig. 4:16 Plantain: the flowers near the top of the inflorescence are in the female condition showing projecting stigmas, those below are in the male condition with projecting anthers.

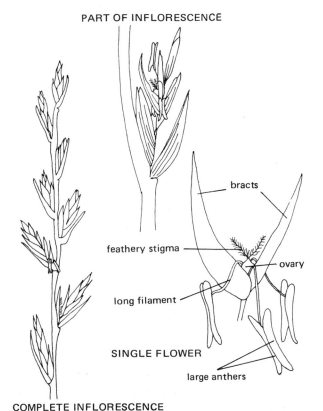

PART OF INFLORESCENCE

bracts

feathery stigma

ovary

long filament

SINGLE FLOWER

large anthers

COMPLETE INFLORESCENCE

2. The stigmas ripen before the stamens, e.g. plantain.
3. The flowers are of more than one kind, the length of the style in each case being different, thus causing the position of the stigma to vary in relation to that of the stamens, e.g. primrose.
4. The flower has a stigma on which its own pollen will not develop, e.g. sweet pea.
5. Some flowers lack stamens, others on the same plant lack carpels, e.g. hazel.
6. All the flowers on some plants lack stamens, on other plants they all lack carpels, e.g. holly.

Anthers, ovules and fertilization

In order to understand how fertilization takes place we must now study the structure of a stamen and ovule and see how the gametes are formed.

Fig. 4:17 (left) Rye grass (*Lolium*) showing flower structure.

50

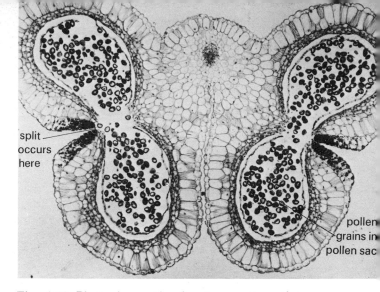

Fig. 4:18 Photomicrograph of a transverse section through the anthers of a stamen showing the four pollen sacs.

Structure of the anther

Each stamen has two anthers; each anther is bi-lobed and within each lobe is a **pollen sac**. So in a transverse section through the anthers (Fig. 4:18) you see four pollen sacs. The pollen grains are formed inside the sacs and when ripe, a longitudinal split occurs between the two pollen sacs of each side and the walls curl back and expose the pollen. The pollen may then be transferred by insects, wind or some other agent during the process of pollination.

Structure of the ovule

The number of ovules formed in the ovary portion of the carpel varies from one in the buttercup to many thousands in some orchids. Each ovule grows from a place inside the ovary called the **placenta**, the position varying according to the species. As the ovule grows it becomes surrounded by a protective coat except for a hole at one end called the **micropyle**. Inside, various nuclei are formed by cell division, one of which enlarges and becomes the female gamete or egg cell.

What happens to the pollen grains when they reach the stigma?

A pollen grain, on reaching the stigma, absorbs sugar and water, swells up and starts to grow by putting out a hair-like tube—the **pollen tube**. This tube acts rather like the hypha of a fungus as it produces enzymes at its tip and dissolves the tissues of the style as it grows, feeding on the products. Eventually the pollen tube travels down the whole length of the style into the ovary, its direction of growth being influenced by a sugar solution which is secreted through the micropyle of an ovule (Fig. 4:20). During the early stages of pollen tube growth several nuclei are formed within it, two of which are male gametes. On reaching the ovule the end of the pollen tube bursts and one male gamete passes into the ovule and fuses with the female gamete. This is fertilization.

You can observe some aspects of this process as follows:

Add a little agar powder to a 10% sucrose solution, warm to dissolve it and then pour a little over the surface of several microscope slides; when cool the solution will set like a jelly. Scatter pollen from different flowers over

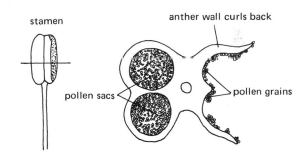

Fig. 4:19 Diagram showing how the anthers split and expose the pollen.

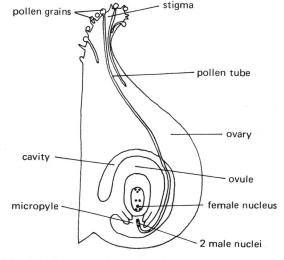

Fig. 4:20 Diagram of a section through a single carpel at fertilization stage.

51

the jelly; place each slide in a Petri dish with enough water just to cover the bottom but not the top of the slide (the slide can be raised slightly with Plasticine). Place the top on the Petri dish and keep at about 20°C. Examine after two or three hours and again the next day. Describe carefully what you see.

In the meantime, remove three or four carpels from a buttercup flower. Choose a flower in which the stamens have curled back to expose the carpels as pollination is likely to have taken place in such a flower. Place one or two carpels on a slide in some water, put another slide on top and squash firmly. Examine under the microscope and look for pollen grains on the stigma and pollen tubes in the cavity of the ovary. Attempt to follow the pollen tubes back through the style to the pollen grains.

What happens after fertilization?

Externally, the flower starts to wither and the sepals, petals and stamens usually fall off, although in some the sepals persist. At the same time the gynoecium enlarges considerably. Inside the ovule, the zygote, formed by the fusion of the gametes, undergoes cell division to form an embryo plant consisting of a miniature shoot (**plumule**), a root (**radicle**) and one or two modified leaves (**cotyledons**). A food store is also laid down either round the embryo or within its cotyledons. This food comes from the green leaves of the plant. As the ovules develop into mature seeds their outer membranes become harder to form the **testa** and finally water is withdrawn from the tissues of the seed. It is now ready for dispersal. At the same time the ovary wall either dries up, or becomes fleshy, according to the species; it becomes the wall or **pericarp** of the fruit. Thus the carpel with its contained ovules becomes the fruit with its seeds.

Types of fruits

A true fruit is formed from the gynoecium *only*, but many common fruits such as apples and strawberries are called false fruits because they have developed from other parts of the flower in addition to the gynoecium. Thus the part of the apple you eat is the receptacle which has become fleshy and the core with its pips is the fruit with its seeds (Fig. 4:21b). In the

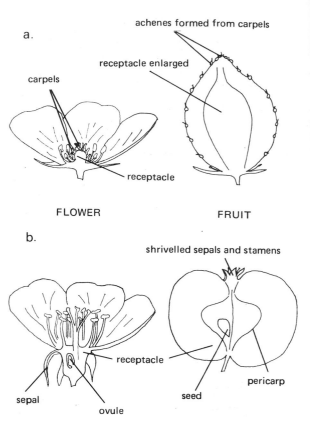

Fig. 4:21 Sectional diagrams showing fruit formation: a) strawberry b) apple.

strawberry, the succulent part is the receptacle and the pips on the surface are the true fruits, each containing one seed (Fig. 4:21a).

True fruits are said to be either **dry** or **succulent** according to whether the pericarp dries up or becomes fleshy. Most dry fruits are **dehiscent**, i.e. their walls split in various ways to let the seeds out; this aids seed dispersal. Other dry fruits are **indehiscent**, i.e. their walls do not split as this is unnecessary as they contain only a single seed. Examples of various kinds of fruits are shown in Fig. 4:22.

Dispersal of fruits and seeds

If all the seeds of a plant fell on to the ground below it and germinated they would compete with each other and the parent for light, water and mineral salts and the majority would die. Plants cannot move of their own accord, but have special adaptations by which their fruits or seeds may be dispersed by some other agent

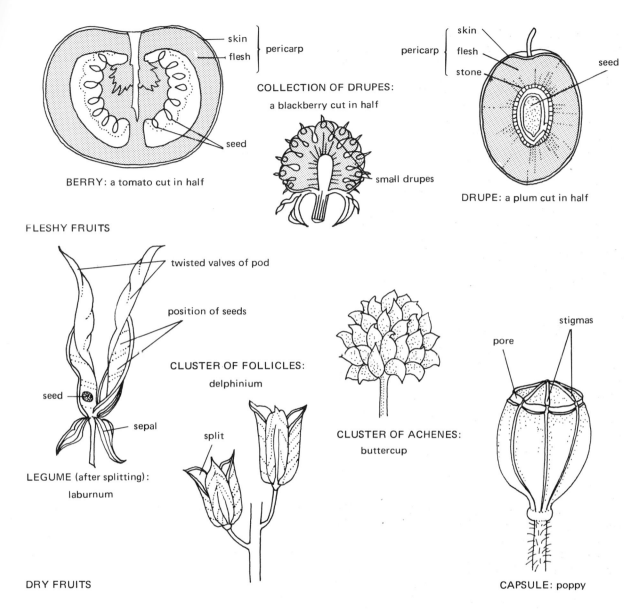

FLESHY FRUITS

DRY FRUITS

Fig. 4:22 Selected fruits

away from the parent. The most usual agents of dispersal are wind and animals, but some aquatic plants use water and a few have mechanical devices to flick the seeds away. Here are some of the methods employed:

Dispersal by wind

To make the best use of wind the fruits or seeds need either to be very small, as in orchids, where they are almost as small as pollen grains and very numerous, or to have a large surface area compared with their volume. The latter is achieved in various ways: some such as elm, ash and sycamore have wing-like extensions, others such as dandelion, thistle and willow-herb have parachutes, while others are very hairy, e.g. clematis (Fig. 4:24).

In the poppy, the ripe fruit forms a ring of holes, through which the tiny seeds can be dispersed. The fruit develops on a long stiff stalk, so when the wind blows it to and fro the seeds are shaken out like pepper out of a pepper pot. This is known as a **censer** mechanism.

53

Dispersal by animals

This may occur as a result of chance contact or because the fruits or seeds are edible (Fig. 4:23). Many plants have dry, hooked fruits which catch in the furry coats of mammals and fall off later. This applies particularly to mammals with curly hair, e.g. some dogs and sheep. Burdock, cleavers and *Geum* may be dispersed in this manner. Others are carried by water birds as they fly from one place to another, the seeds being lodged in the mud which clings to their feet.

Dry edible fruits such as acorns and hazelnuts are collected by squirrels and jays and buried as a food store; if they are not found again they may germinate.

Succulent fruits may be dispersed as a result of being eaten by birds and mammals. They may be eaten whole and the seeds passed through the gut to be deposited with the faeces as in currants and holly, or they may be partially eaten and the seeds scattered as in plum, cherry and mistletoe. In the mistletoe the seed is sticky and adheres to the beak of the thrush which may press it into a crevice in the bark of a tree in order to wipe it off its beak. Here it germinates and the seedling becomes parasitic on the tree.

Dispersal by water

For this to be effective the fruit or seed must be buoyant enough to float. In coconuts this is achieved by air trapped in the outer covering (Fig. 4:24).

Self dispersal mechanisms

Plants which flick their seeds out do so as a result of the uneven drying of their fruits. For example, when the pods of gorse, broom and lupin dry, tensions are set up in their walls which cause the pod to burst suddenly. When this happens the valves of the pod twist and the seeds are thrown out with considerable force; sometimes they travel several yards. Balsam also has an 'explosive' fruit (Fig. 4:24).

Collect as many kinds of fruits as you can and classify them into their main groups (see table p. 56) and note any adaptations that they have for dispersal.

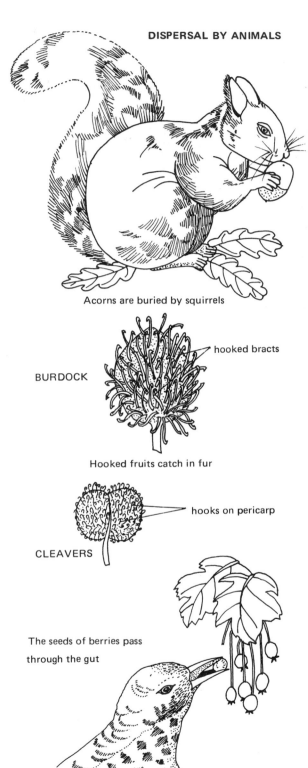

Acorns are buried by squirrels

BURDOCK — hooked bracts

Hooked fruits catch in fur

CLEAVERS — hooks on pericarp

The seeds of berries pass through the gut

Fig. 4:23

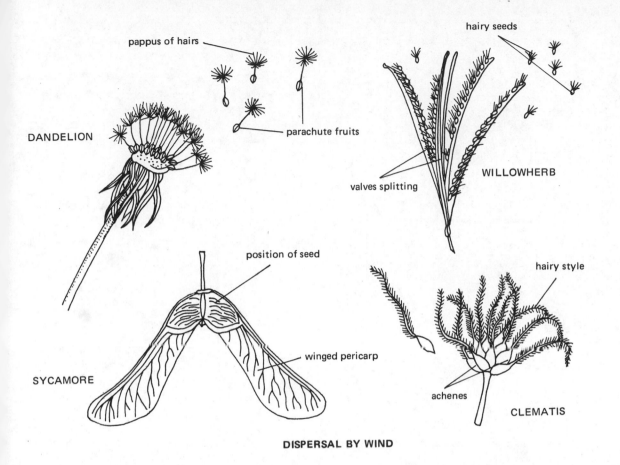

DANDELION

pappus of hairs

parachute fruits

hairy seeds

WILLOWHERB

valves splitting

position of seed

winged pericarp

SYCAMORE

hairy style

achenes

CLEMATIS

DISPERSAL BY WIND

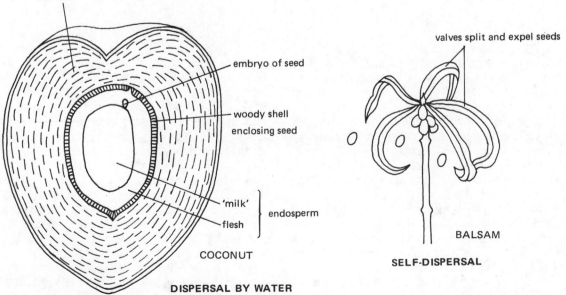

air trapped in fibrous coat helps it to float

embryo of seed

woody shell
enclosing seed

'milk'

flesh

} endosperm

COCONUT

DISPERSAL BY WATER

valves split and expel seeds

BALSAM

SELF-DISPERSAL

Fig. 4:24

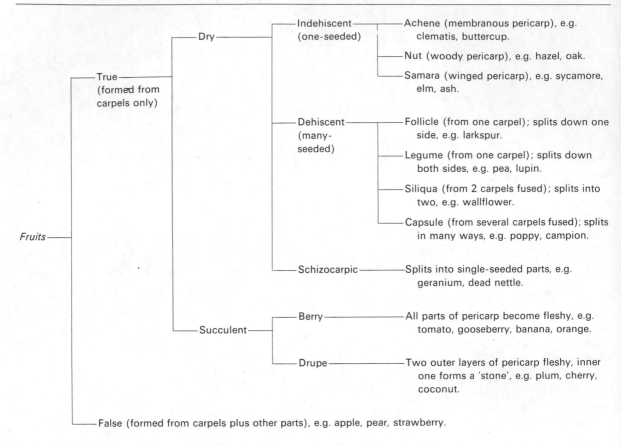

Fruits
├─ True (formed from carpels only)
│   ├─ Dry
│   │   ├─ Indehiscent (one-seeded)
│   │   │   ├─ Achene (membranous pericarp), e.g. clematis, buttercup.
│   │   │   ├─ Nut (woody pericarp), e.g. hazel, oak.
│   │   │   └─ Samara (winged pericarp), e.g. sycamore, elm, ash.
│   │   ├─ Dehiscent (many-seeded)
│   │   │   ├─ Follicle (from one carpel); splits down one side, e.g. larkspur.
│   │   │   ├─ Legume (from one carpel); splits down both sides, e.g. pea, lupin.
│   │   │   ├─ Siliqua (from 2 carpels fused); splits into two, e.g. wallflower.
│   │   │   └─ Capsule (from several carpels fused); splits in many ways, e.g. poppy, campion.
│   │   └─ Schizocarpic ─ Splits into single-seeded parts, e.g. geranium, dead nettle.
│   └─ Succulent
│   ├─ Berry ─ All parts of pericarp become fleshy, e.g. tomato, gooseberry, banana, orange.
│   └─ Drupe ─ Two outer layers of pericarp fleshy, inner one forms a 'stone', e.g. plum, cherry, coconut.
└─ False (formed from carpels plus other parts), e.g. apple, pear, strawberry.

SIMPLIFIED CLASSIFICATION OF FRUITS

Comparison of vegetative and sexual reproduction in flowering plants

We have seen how flowering plants may use two very different methods of reproduction, an asexual or vegetative method involving a process of budding and a sexual method resulting in the formation of seeds. Both methods have advantages and disadvantages.

The vegetative method is a relatively safe one; food is available from the parent plant for early development, the progeny can be quite large before becoming independent and so they can compete better with other plants in the neighbourhood for survival. This method results in the species growing in clumps or patches and in consequence competition between plants of the same species may increase, but this disadvantage may well be offset by the effect of crowding out plants of other species and the protection afforded to its own. These clumps are also very resistant to unfavourable external conditions and if the upper parts are destroyed they can form new aerial shoots from their underground buds to replace them. The chief disadvantage is that dispersal is limited to the immediate vicinity of the plant and the method is usually slow. There is also the fact that the daughter plants will inherit the exact characteristics of the parent, in other words, they will vary very little (Ch. 14). This may be an advantage in the short term, but not if the habitat changes.

The sexual method is far more hazardous as most pollen does not reach a suitable stigma and the majority of seeds never reach a suitable habitat; vast numbers are eaten and if they do germinate they must quickly form leaves and reach the light to form food before their small stores of food are used up. They may also have to suffer fierce competition from other plants. For these reasons large number of seeds have to be produced in order for some to survive. On the other hand, if conditions are favourable, this method is excellent and vast numbers of new plants may be produced. A great advantage of the sexual method is that seeds can be taken great distances and so dispersal of the species is much more effective. If disaster comes to one area it may be recolonised more easily from another. Also, sexual reproduction allows variation in the progeny and in the long term this is a great advantage if conditions should change. A further advantage is that seeds can survive over long periods of unfavourable conditions in a state of dormancy.

Thus both methods have their advantages and disadvantages and it is not surprising that many plants obtain the best of both worlds by using both methods.

5

Germination and growth in plants

Seed structure

Seeds have a protective covering, the testa, which encloses an embryo plant consisting of a shoot—the plumule, a root—the radicle, one or two modified leaves—the cotyledons, and a food store which either surrounds the embryo and is called **endosperm** or is stored inside the cotyledons.

Examine broad beans, french beans and maize which have been soaked in water to soften them. Strictly speaking, a maize grain is a fruit containing one seed, the testa having fused with the pericarp. Compare their external features and note similarities and differences. Squeeze both beans gently; you should see the position of the micropyle by the bubble which is formed at that point.

Remove the testa from the two beans and open out the cotyledons to display the plumule. Where is the food stored? What is it composed of? Carry out tests for starch, sugar and protein. (Book 1 p. 161).

Now cut the maize longitudinally, slightly to one side of the mid-line (Fig. 5:3). Note the size of the embryo and the position of the food store; also find out the nature of the food.

Germination

You can compare the method of germination in these three species by placing one seed of each in a glass jar, kept in position by wet blotting paper as in Fig. 5:1. Over the next fortnight note how the plumule and radicle emerge and observe what happens to the cotyledons in each case. Although the broad and french bean seeds are very similar in structure you will notice an important difference in their development. When cotyledons remain in the seed, the seedling is said to be **hypogeal**; if the cotyledons come above ground and turn green, the seedling is **epigeal**.

Conditions necessary for germination

When you try to grow seeds in the garden they do not always come up as well as you had hoped; sometimes, they may not germinate at all. There are many possible reasons for this. First let us find out the importance of such factors as oxygen, water, temperature and light on germination.

Set up 5 large specimen tubes with cotton wool pressed into the bottom of each. Label them 1–5 (Fig. 5:4). Put 10 mustard seeds in each. Add enough water to soak the cotton wool in all tubes except tube 3. Remove oxygen from tube 4 by suspending a small tube of pyrogallol (an oxygen absorber) by means of a cotton thread and closing with a cork.

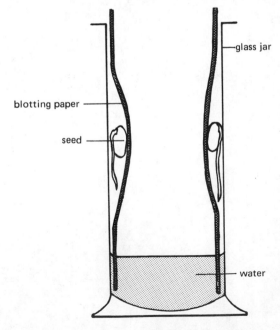

Fig. 5:1 Apparatus for observing the stages of seed germination.

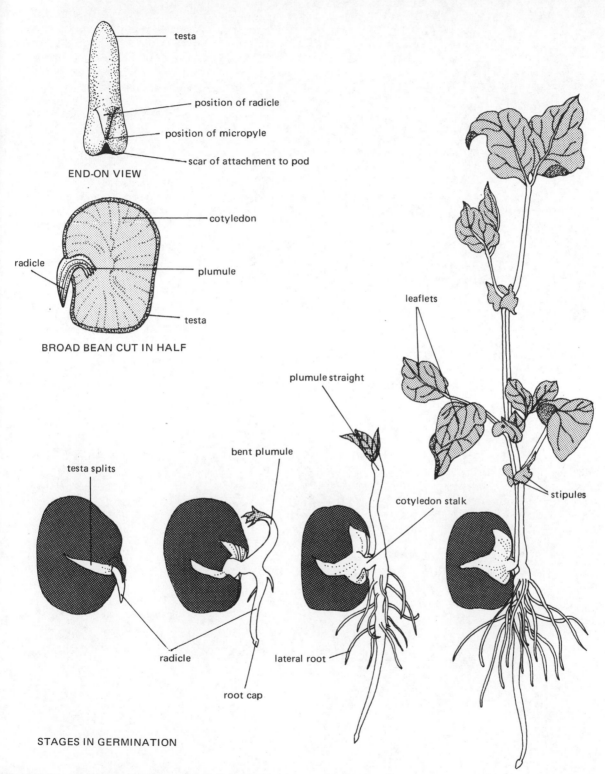

testa

END-ON VIEW

position of radicle

position of micropyle

scar of attachment to pod

cotyledon

radicle

plumule

testa

BROAD BEAN CUT IN HALF

leaflets

plumule straight

cotyledon stalk

stipules

testa splits

bent plumule

radicle

root cap

lateral root

STAGES IN GERMINATION

Fig. 5:2 Structure and germination of the broad bean.

59

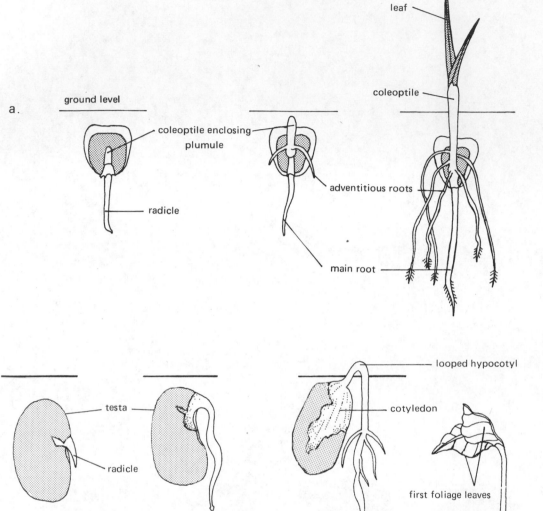

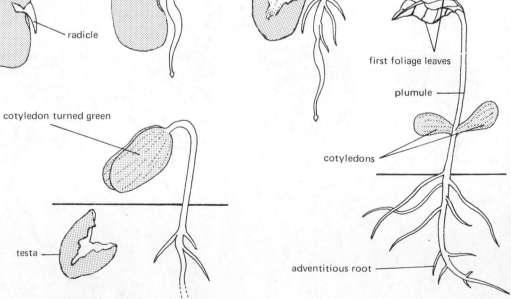

Fig. 5:3 Stages in the germination of a) maize b) french bean.

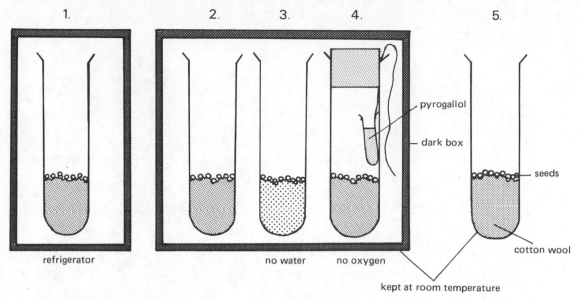

Fig. 5:4 Apparatus for determining the conditions necessary for germination.

Put tubes 2, 3 and 4 in a dark box and 5 in the light. Keep all these at room temperature. Put 1 in a refrigerator (which is also, of course, dark).

Examine the tubes periodically. What are your conclusions concerning the factors necessary for germination?

Comparisons of 1 and 2 will show the effect of temperature.

Comparisons of 2 and 3 will show the effect of the absence of water.

Comparisons of 2 and 4 will show the effect of the absence of oxygen.

Comparisons of 2 and 5 will show the effect of the absence of light.

We can understand the importance of these factors if we know what is happening inside the seed when it germinates.

First, water is absorbed by the seed, mainly through the micropyle, by the physical process of imbibition (soaking in). This stimulates the cytoplasm of the cells to secrete enzymes which turn the stored food into soluble substances, including sugar, which are then available for growth. Oxygen is used to respire some of this sugar to provide the energy for growth. These reactions will only take place at a significant speed if a suitable temperature is reached (this varies for different species).

Light is not essential for germination, but when the seedling comes up above the ground light is necessary for the formation of chlorophyll and consequently for the formation of more food. It also helps the young plant to grow strong. In addition, light influences the direction of growth (p. 67).

Growth in length

Growth in length of both radicle and plumule is rapid, but uneven. You can see for yourself the regions of growth by using the broad bean seedlings you have been growing in the glass jar or by germinating some more.

Remove a few beans with radicles about 3cm in length and mark each radicle at 2mm intervals with a cotton soaked in Indian ink (Fig. 5:5). Make sure the radicles are not wet beforehand otherwise the ink may smudge. Replace the beans when the ink is dry and examine after a day or so. Which parts have grown most?

Growth in length is the result of two processes, cell division and vacuolation of the cells. Cell division takes place at the tips of all roots and shoots, which are thus called **growing points**. Behind these regions the cells

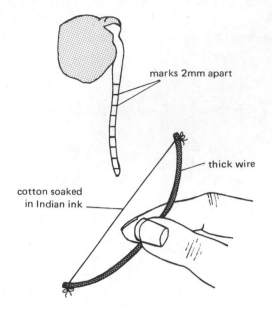

marks 2mm apart

thick wire

cotton soaked
in Indian ink

Fig. 5:5 Method of marking a bean radicle.

absorb water and become vacuolated, with the
result that they elongate considerably, causing
the growth in length. Further back the cells
stop growing and become differentiated into
their permanent form.

When cells divide it is always the nucleus that
plays the essential part. The nucleus divides
first by a process called **mitosis** which is
basically similar in both plants and animals.
The cytoplasm divides afterwards.

Mitosis

This process of cell division which is a characteristic of growth involves thread-like structures called **chromosomes**. All but the most
primitive species of plant and animal have a
definite number of pairs of chromosomes in
each of their nuclei and during mitosis this
number is exactly maintained. Thus if a cell
has 23 pairs, as in man, when it has divided
into two, 23 pairs still remain in each. This is
accomplished by an exact duplication of each
chromosome so that two new chromosomes
are formed, one of which goes to one daughter
cell and the other to the second. The significance of this exact duplication is discussed on
p. 170.

Study the stages of mitosis in Fig. 5:7 to
see the details of the process. Note that when

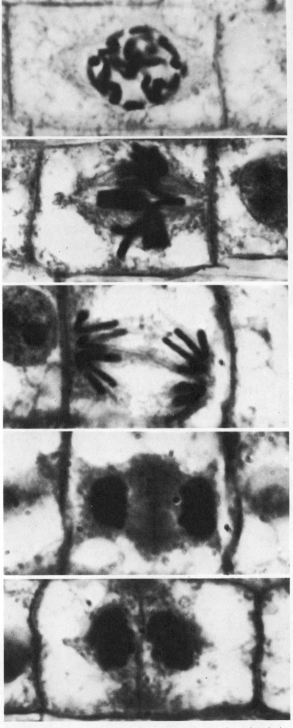

Fig. 5:6 Photomicrographs showing stages in mitosis in
cells from the root tip of an onion. Compare with
Figure 5:7.

62

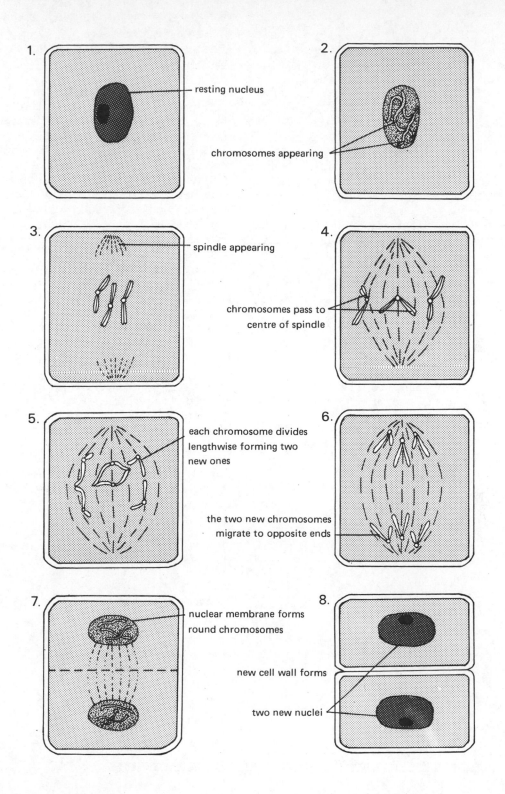

Fig. 5:7 Mitosis in a plant cell (for the sake of clarity only 3 chromosomes are shown).

the nucleus has divided into two, in plant cells a new cell wall is formed between them by the cytoplasm. In the division of animal cells the outer cell membrane constricts and then completely separates into two cells.

You can see some of the stages for yourself by looking at specially prepared longitudinal sections through the root tip of an onion (Fig. 5:6). You can also make a preparation yourself by separating the individual cells by a squash technique and then staining the nuclei:

Cut off the last 4mm of an onion root. Place it in a test tube containing a little Molar hydrochloric acid at about 60°C for 5 minutes. This treatment will soften the tissues.

Decant the liquid carefully. Wash the root tip in a watch glass containing water to remove the acid.

Place the tip on a slide, remove excess water with filter paper and place two small drops of acetic orcein stain on it.

Now gently tap the root with the end of a glass rod or scalpel handle until it is completely broken up. Place a coverslip on top and warm it *gently* for a few seconds over a very low flame; it must not boil.

Put a piece of filter paper over the preparation and press the coverslip with your thumb, avoiding any sideways movement.

Examine the individual cells under the high power and find as many stages of mitosis as you can.

Measurement of rate of growth in length

The rate of growth varies with temperature, the amount of water and nutrient present in the soil, and the amount of light the plant receives. This can be demonstrated by using an **auxonometer** (Fig. 5:8) which automatically records the growth of a shoot. The growth will be magnified according to the ratio between the two arms of the lever. The plant should be kept well watered. Cotton wool should be used at the place where the cotton is fastened to prevent damage to the delicate tip. The cotton should be short so that if the humidity of the air varies any error due to changes in length of the cotton will be negligible. The compensatory weight should be just heavy enough to prevent the lever from pulling on the plant. The smoked disc is rotated by an electric motor at a speed of one revolution per hour. As the disc rotates, a bristle attached to the end of the bar makes a fine tracing on its smoked surface in the form of a continuous spiral. The distance between adjacent lines is proportional to the amount of growth that has taken place.

This apparatus, if examined after 24 hours, will indicate the effect of light and darkness on the growth rate if other factors such as temperature are kept constant. How could the apparatus be used to investigate the effect of temperature on growth?

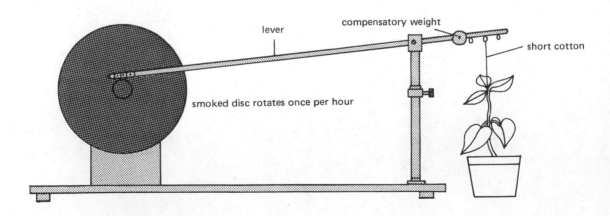

Fig. 5:8 Auxonometer.

Growth in thickness

As we have seen, growth in *length* occurs as a result of cell division and elongation at the tips of roots and shoots. Growth in thickness also results from cell division, but those concerned are the cambium cells which lie between the xylem and phloem. These cells divide in such a way that they form new xylem towards the inside and more phloem towards the outside. In very young stems the cambium is restricted to a position within the vascular bundles, but later it joins up to form a complete ring (Fig. 5:9). In woody plants, such as trees, this formation of secondary xylem and phloem continues throughout life, causing the trunk to get thicker and thicker. So a section through a 3-year old stem (Fig. 5:10) shows 3 rings of secondary xylem making up the wood. New phloem is added each year, but this growth is not indicated by rings. The phloem lies just under the bark.

The rings in the xylem occur because growth is not regular. In the late summer and early autumn only small vessels and fibres are produced, in the late autumn and winter growth ceases and in the spring much larger vessels are formed. Thus the small autumn vessels are found next to the large spring vessels and their difference in texture is visible as a ring. As these rings occur annually the age of a particular branch or trunk can be determined by counting them.

Bark

As the stem thickens from within, the epidermis becomes stretched and eventually splits and peels off. Its protective function is taken over by a layer of cork formed from a ring of **cork cambium** which usually arises in the outer cortex. This layer of cork becomes thicker with repeated division of the cork cambium and becomes the **bark**. Cork is non-living and is impermeable to water and gases, but the growing stem still requires oxygen for respiration. This is obtained through **lenticels** which are pores in the bark, where the corky cells are loose and allow gases to diffuse between them (Fig. 5:11).

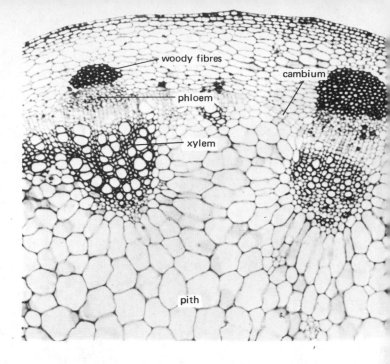

Fig. 5:9 (above) Photomicrograph of a transverse section of a sunflower stem showing the cambium as a continuous ring.

Fig. 5:10 (below) Photomicrograph of a 3-year old stem of elm.

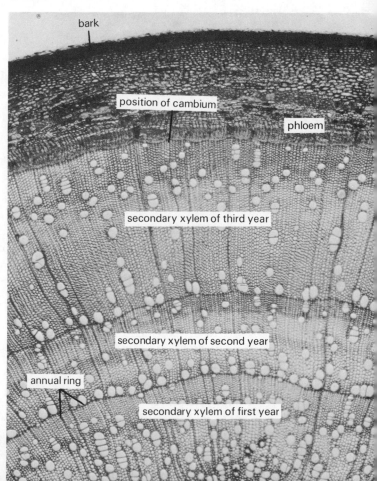

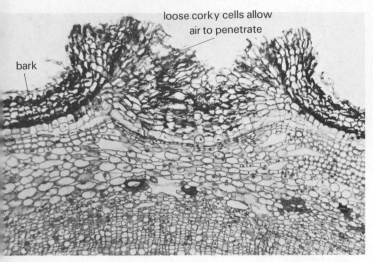

Fig. 5:11 Photomicrograph of a section through a lenticel of elder.

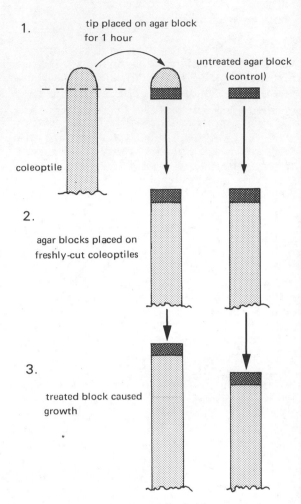

Fig. 5:12 The type of experiment used to demonstrate how auxin in the tip of the coleoptile (plumule sheath) controls growth.

Control of growth

We saw in Chapter 3 that growth in animals was controlled by hormones secreted into the blood stream from the thyroid and pituitary glands. Are there any comparable substances in plants? First, consider this classical experiment:

A number of oat seedlings were grown in the dark and after a few days the tips of several coleoptiles were cut off and placed upright on tiny blocks of agar for one hour. Then further coleoptiles were decapitated and the agar blocks were carefully placed on the cut ends of half of them, while untreated agar blocks were placed on the others to act as controls. Within a matter of a few hours under warm conditions those with treated agar had grown considerably, while those with untreated agar had grown hardly at all (Fig. 5:12).

From this and other rather similar experiments it was concluded that the tip of the shoot produced a substance, soluble in water, which diffused backwards into the region of the stem behind the tip causing it to elongate. The agar, by absorbing this substance, acted in the same way as the tip when placed on the cut end.

This growth-regulating substance, first called **auxin**, was analysed and found to be indolyl acetic acid (IAA). It is now known to regulate the vacuolation of cells and thus control their growth in length. This is just one of the plant hormones.

What controls the direction of growth?

When you were growing seedlings, you will have noticed that whatever the position of the seed in the jar, the radicle grew downwards and the shoot upwards. Also, when plants are placed on a window ledge the shoots tend to bend towards the light, and when you grow runner beans in the garden they twine up the supports in a definite manner. What controls all these directional growth movements? Clearly the plant organ in each case is responding to some stimulus such as gravity or light or touch.

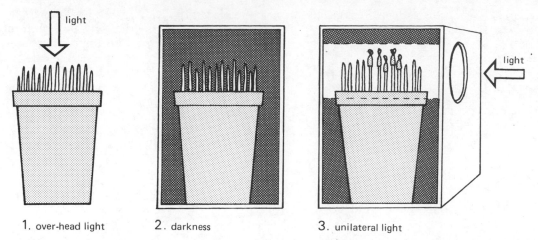

| 1. over-head light | 2. darkness | 3. unilateral light |

Fig. 5:13 Apparatus for determining the effect of light on the directional growth of seedlings.

Phototropism

This is the growth response of a plant organ to the stimulus of light; it is a directional bending in relation to the position of the light source.

Place about 10 oat grains in each of 3 small pots containing damp sawdust, planting them just under the surface. Let the seeds germinate in the dark by keeping them in an incubator at about 25°C for 4 or 5 days.

Make some small aluminium foil caps by moulding them on a matchstick. Place them over half the seedlings in one of the pots. The caps will keep the tips in darkness. Test their reaction to directional light by arranging them in boxes as shown in Fig. 5:13. Examine the seedlings after a few days. How has the direction of the light source affected the bending of the shoots? Which part of the shoot is most sensitive to light? In which region of the shoot has the bending occurred? Which set has grown the most?

This bending of a plant shoot towards the light source is called **positive phototropism**. The majority of *roots* show no reaction to light although a few such as mustard do bend away from a light source (they are said to be **negatively phototropic**). Leaves usually respond so that their surfaces are at right angles to the light. In woods, with the light coming mainly from above, the leaves of many trees such as beech grow in such a way that the lower ones find the gaps between those above them, so that they are exposed to the maximum light available. This arrangement is known as a **leaf mosaic**.

Geotropism

This is the response of a plant organ to the stimulus of gravity. As you have seen from growing seedlings, shoots grow upwards against gravity and roots grow downwards. How can you prove that this really is due to gravitational force? To do this you need to subject some plants to gravity and others to none. The latter would only be possible if the plants were grown in a space module when gravitational forces were not acting! Which way do you think the roots and shoots would grow if this were done?

It is not possible to eliminate the force of gravity under normal laboratory conditions, but it is possible to even out its effect by rotating the seedlings slowly and regularly so that all regions receive the same force. This can be demonstrated by using a **klinostat** (Fig. 5:14). An electric motor causes the cork disc to rotate about once every 15 minutes. The seedlings are pinned to the cork which is covered with soaking wet cotton wool. They are arranged so that the roots point in various directions. A celluloid cover is placed on top to keep the air inside saturated with water and so prevent the seedlings from drying up. A similar apparatus should be set up which is

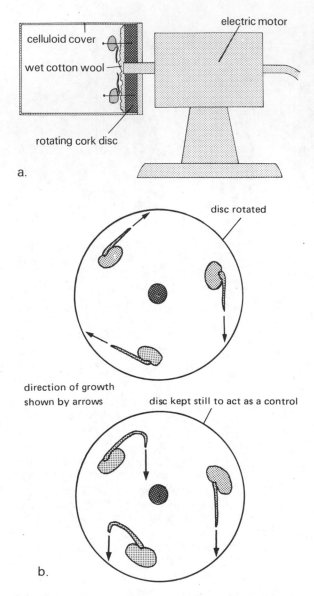

Fig. 5:14 Experiment demonstrating the effect of gravity on the directional growth of seedlings: a) klinostat b) cork discs seen from in front.

not rotated, to act as a control. After about 2 days the roots in the control should all have grown downwards and those which were rotated should have gone on growing in the direction in which they were growing before.

Which part of the root is sensitive?

Take two broad bean seedlings which have been grown in damp sawdust and which have straight radicles about 2cm long. Mark the radicles with Indian ink at regular intervals as you did in a previous experiment, and then cut off the tip of one of them. Place them on wet cotton wool in a Petri dish as in Fig. 5:15. Now place the dish in a dark place (why?) on its side in such a position that both radicles are horizontal. Examine each day. Which part of the root is sensitive to the stimulus of gravity? Which part of the root causes the bending?

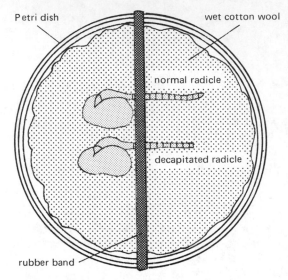

Fig. 5:15 Experiment to find out which part of a root is sensitive to gravity.

The mechanism of curvature

We should have seen from these experiments involving the effect of both light and gravity that it is the tip of the root or shoot which is sensitive to this stimulus, but it is the region behind it—the region of vacuolation—that responds. It is also a fact that when a root or shoot bends it is because the cells on one side vacuolate more than on the other, thus extending them more. Could it be that in phototropic or geotropic bending more IAA reaches one side than the other?

This possibility was tested by using oat coleoptiles, as in the auxin experiment, (p. 66), by cutting off their tips and replacing them at once eccentrically as in Fig. 5:16. After a few hours the coleoptiles had bent. It was also found that if the tips were placed on agar, so that the IAA could diffuse into it, and the agar block was put back eccentrically, the same

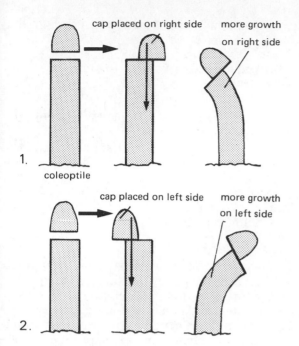

1.

coleoptile

cap placed on right side

more growth on right side

2.

cap placed on left side

more growth on left side

Fig. 5:16 The type of experiment used to demonstrate the effect of auxin on stem curvature. The vertical black arrows indicate the movement of auxin.

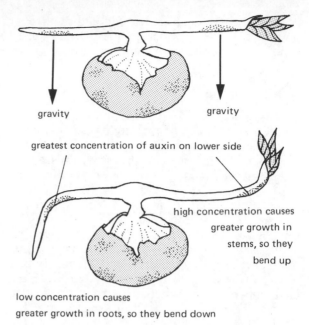

gravity   gravity

greatest concentration of auxin on lower side

high concentration causes greater growth in stems, so they bend up

low concentration causes greater growth in roots, so they bend down

Fig. 5:18 The effect of gravity on root and shoot.

thing happened. So IAA can cause bending if there is a greater concentration of it on one side of a plant organ. IAA is such a powerful growth regulator that one part in a million parts of water is enough to cause a growth curvature, but it has been shown that different organs respond differently to various concentrations. Consider the effect of gravity on root and shoot. When a seedling is placed horizontally (Fig. 5:18) the auxin tends to move towards the lower side of both root and shoot, but the root bends down and the shoot up. Why should they not both bend in the same direction? It is now known that although the same substance is present in both root and shoot a higher concentration of IAA causes elongation in the stems, but in the root it is a lower concentration that has the greater effect.

If too high a concentration of auxin is given, the plant will be killed. This is the principle behind the use of **selective weedkillers**. These are synthetic substances which act like plant hormones. When used in higher concentration different species of plants show varying tolerance to them, so they kill off some and not others. One substance known as **2.4-D** is widely used for spraying on lawns. It is absorbed through the surface of the leaves but affects monocotyledons such as grasses far less than dicotyledons such as dandelions and daisies which quickly die as a result. It is also used for spraying wheat fields to eliminate dicotyledonous weeds which would compete for soil nutrients. (See also Fig. 13:8.)

Hormones also control the formation of adventitious roots; hence the practice of dipping a cutting in a hormone powder before planting it. This has enabled people to propagate plants which otherwise do not readily form adventitious roots.

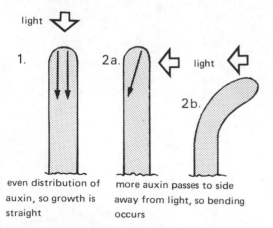

light

1.   2a.   light

2b.

even distribution of auxin, so growth is straight

more auxin passes to side away from light, so bending occurs

Fig. 5:17 The effect of light on the distribution of auxin.

Hormones also affect leaf fall and the dropping of fruit. By spraying an apple crop with IAA it is possible to prevent the fruit from falling off prematurely.

Other tropisms

In this chapter we have concentrated attention on the effects of gravity and light on directional growth and the hormones which help to bring this about. These are not the only tropisms.

Haptotropism is the growth bending resulting from a touch stimulus. Many plants have tendrils which are used for support. If you stroke the inside of a tendril with a matchstick it will soon begin to curve. Other plants such as runner beans and bindweed have stems which are sensitive, and in consequence twine round any support they touch. **Hydrotropism** is shown by roots when they grow towards a source of water; this is easily seen along the banks of a river where the tree roots are often visible.

We have already described an example of **chemotropism** (p. 51) where the direction of growth of the pollen tube is controlled by sugar secreted through the micropyle of the ovule.

All these tropisms are caused by greater growth on one side of a plant organ than on the other.

Comparison of growth in plants and animals

Similarities
1. Growth in both is brought about by mitotic division of cells.
2. Its rate and extent is modified by such factors as temperature and the availability of suitable nutrients, although the general pattern is determined by hereditary factors.
3. Hormones play an important part in controlling growth.

Differences
1. Growth in both is continuous throughout the life of the plant, but animals stop growing after a time, although cell division continues for repair and replacement.
2. Growth is restricted to special growing points in plants, but growth in animals occurs throughout their bodies.
3. As a result of 2, plants constantly change their shape, e.g. as buds put out new shoots, but animals keep roughly the same shape once their developmental stages have taken place.

70

6

Keeping the body in a steady state

If you can imagine an organism living in a habitat where conditions never altered and where it was perfectly adapted to those conditions, there would be a perfect balance between that organism and its environment. In nature this never happens. Instead, we find many varieties of habitats and many changes of conditions within them, and in order for the organism to survive there must be many regulatory mechanisms which help to keep a state of balance when conditions change. The term used to describe this 'steady state' is **homeostasis**. In this chapter we shall examine some of the ways in which this balance is maintained.

Oxygen/carbon dioxide balance

One example of homeostasis we have already come across is the maintenance of the correct amount of oxygen within our body. When we take vigorous exercise much more oxygen is used by the muscles and so the amount of carbon dioxide in the blood increases. The brain responds to the higher level of carbon dioxide by increasing the rate at which nerve impulses are sent to the heart and respiratory muscles; this results in a faster heart beat and quicker and deeper breathing, and eventually the oxygen/carbon dioxide balance is restored.

Just as an organism has to maintain a balance with the *external* environment, so the living tissues have to maintain a balance with the body fluids which constitute its *internal* environment.

We have already seen (Book 1 p. 125) how very sensitive blood corpuscles are to osmotic changes in the plasma. The problem of maintaining blood in a steady state is not simple because its composition is changing all the time. When blood passes through the capillaries it loses oxygen, sugar and other nutrients to the cells and at the same time gains carbon dioxide and other products of cell metabolism. The amount of water in the blood is also fluctuating. Let us now consider some of these problems in more detail.

The problems of water balance

The control of the water content of the body is known as **osmo-regulation**. Such a mechanism is essential because if too much water is absorbed into the blood its osmotic strength would become less than that of the surrounding body tissues, and water would be drawn into them causing them to swell up. Conversely, if the blood became too concentrated, water would be drawn out of the tissues and they would shrink.

The problem, therefore, is to ensure that the water lost by the body is balanced by what is taken in. In an average person the overall equation would be something like this:

$$\text{water drunk } (1600\text{cm}^3) \atop \text{water in food } (900\text{cm}^3) \atop \text{water from tissue respiration } (400\text{cm}^3) = \text{water in urine } (1650\text{cm}^3) \atop \text{water in sweat } (650\text{cm}^3) \atop \text{water in faeces } (100\text{cm}^3) \atop \text{water from lungs } (500\text{cm}^3)$$

i.e. 2900cm^3 gained $= 2900\text{cm}^3$ lost

The control mechanisms ensure that if one of these factors alters—for example, when we drink a lot of water—there is a compensatory increase in the water lost in urine. Similarly, on a cold day there is less evaporation through sweating, but more urine is produced.

The problem of nitrogenous waste

When proteins are digested they are broken down into their constituent amino acids, which are then absorbed into the blood stream. Some are used by the cells in the normal processes of growth, i.e. they are built up into proteins again, but the body is incapable of storing excess amino acids. Each amino acid contains nitrogen as part of an amino group $(-NH_2)$. For example, the simplest amino acid has the following formula:

$$\underset{H}{\overset{H}{\underset{|}{N}}}-\underset{H}{\overset{H}{\underset{|}{C}}}-\overset{O}{\overset{\parallel}{C}}OH$$

or more simply $CH_2.NH_2.COOH$

In the liver, amino acids not immediately required by the body are broken down through a process called **deamination**. The amino group is split off and converted into a soluble and relatively harmless substance, **urea**, which is then carried away by the blood; the organic acid which remains can be used as a source of energy in respiration. Other products of protein breakdown include ammonia and uric acid; their quantity varies in different vertebrates. All these substances are collectively called **nitrogenous waste**. These products, if allowed to build up too high a level, would quickly poison the body; their removal or **excretion** is therefore of vital importance. Now let us see how these two processes, excretion of nitrogenous waste and osmo-regulation are carried out by the kidneys.

The urinary system

In Fig. 6:1 you will see that the blood flows into the kidneys via the two renal arteries. As these are short, large, and close to the dorsal aorta, does this suggest anything to you about the pressure of the blood entering the kidneys? Blood is removed from the kidneys via the large renal veins.

We can best think of the kidney as an organ which filters off some of the blood and then puts practically all of it back again, except for a small amount of fluid which passes down the two **ureters** as urine. The urine is temporarily stored in the **bladder**. There are two sets of circular sphincter muscles in the bladder. When the bladder is filling up both these muscles are constricted, so the exit is closed; however, as the pressure of the urine increases, the walls of the bladder are stretched and this triggers off an automatic reflex action which causes the upper sphincter to relax. But the

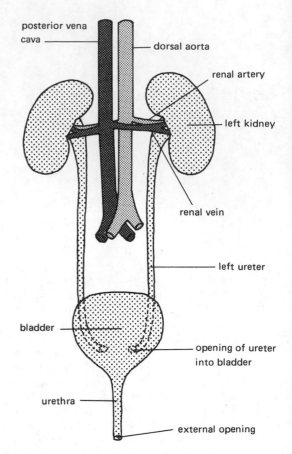

Fig. 6:1 Generalized diagram of the human urinary system.

lower sphincter, in contrast, is under the control of the will, and so urine can still be retained until this muscle is relaxed too. Control of urination is not possessed by very young children but is gradually learnt.

The general arrangement of the urinary system can best be seen by studying a specially prepared dissection of a rabbit or rat.

The kidney

Examine a fresh lamb's kidney. Cut it longitudinally into two equal halves (as in Fig. 6:2) using a razor or sharp scalpel. Compare your specimen with the diagram. What structural differences can you detect between the outer layer or **cortex** and the inner region or **medulla**? You may also be able to see large spaces, the **pelvis**, which lead into the ureter.

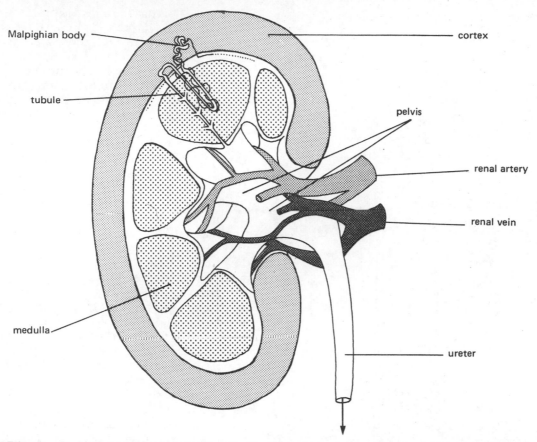

Fig. 6:2 Diagram of a human kidney cut in half longitudinally. A single much enlarged tubule is included to show its position.

Fig. 6:3 Photomicrograph of a longitudinal section through a mammalian kidney.

Some further information can be obtained if you study Fig. 6:3. This is a photomicrograph of a longitudinal section through part of a kidney which has had its larger blood vessels injected with a coloured fluid to show them up better. They appear black in the photograph.

The blobs in this photograph are small knots of arterioles called **glomeruli** (sing. glomerulus). Each one is surrounded by a **Bowman's capsule** which is the cup-like blind end of a single kidney tubule or **nephron** (Fig. 6:4). One human kidney contains about a million nephrons, each approximately 35mm in length. A Bowman's capsule and glomerulus together are called a **Malpighian body**.

What is the significance of all these glomeruli in the cortex? In Fig. 6:4 you will see that the diameter of the walls of the blood vessel entering the glomerulus is greater than that

73

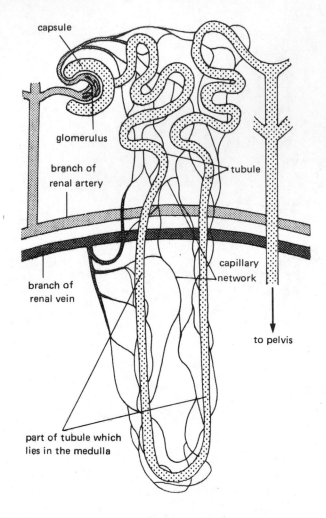

capsule

glomerulus

branch of
renal artery

tubule

branch of
renal vein

capillary
network

to pelvis

part of tubule which
lies in the medulla

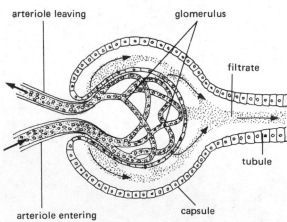

arteriole leaving

glomerulus

filtrate

tubule

arteriole entering

capsule

Fig. 6:4 (above) A single Malpighian body and nephron
showing the blood supply. (below) A single Malpighian
body enlarged.

leaving it. This has the effect of building up the blood pressure within the glomerulus, causing fluid to be forced through the walls of its blood vessels into the capsule and so into the tubule. (The capsule has extremely thin walls which are in close proximity to the glomerulus.) The mechanism by which the fluid is forced out of the blood into the capsule is known as **pressure filtration**.

It has been calculated that every 24 hours an adult human filters off from the blood 170 litres of glomerular fluid—enough to fill the petrol tanks of five or six medium-sized cars! However, from this vast amount only 1.5 litres of urine are produced. In other words about 99% of the filtrate is re-absorbed and passed into the blood as the fluid passes down the tubules.

By the ingenious technique of using ultra-fine glass pipettes, it has been found possible to extract samples of glomerular fluid for analysis. This has enabled a comparison to be made between the composition of glomerular filtrate and that of both blood plasma and urine. Such analyses show that although blood corpuscles and plasma proteins are unable to pass through the walls of the glomerulus, useful substances such as glucose, amino acids and salts do so, as well as some nitrogenous waste. But analysis of urine shows that no glucose or amino acids are normally present and the proportion of salts is very much less than in glomerular filtrate; urea, on the other hand, is found in larger amounts. So it follows from these comparisons that during the passage down the tubules all the sugars and amino acids, most of the salts and 99% of the water are re-absorbed by the cells of the tubules and passed back into the blood capillaries which surround the tubule walls. However, most of the nitrogenous waste, mainly in the form of urea, is selectively retained in the urine.

In this manner the kidney acts as an efficient excretory organ as the most useful substances are retained and the toxic substances are largely eliminated.

The kidney as a homeostatic organ

The kidney is not only an excretory organ, it also *regulates* both the amount of water and urea in the blood. When we drink an excessive amount of water, much dilute urine is produced. Conversely, if we lose a lot of water as sweat and do not have anything to drink, small amounts of more concentrated urine are produced. It seems likely that this difference is related to the way in which the kidney tubules re-absorb the glomerular filtrate. Most re-absorption takes place in the first part of the tubule quite automatically, but there is a means of adjusting how much of the remaining water is absorbed according to conditions in the blood. This 'fine adjustment' takes place both in the second convolution and in the collecting duct (Fig. 6:4). The amount of water in the blood is monitored by special sense receptors in the brain. If the blood becomes too concentrated, i.e. its osmotic pressure is increased, these receptors stimulate the secretion of a hormone from the pituitary gland (p. 25) which, on reaching the kidney via the blood stream, causes the walls of the second convolution and collecting duct to become permeable to water. This means that most of the fluid entering this part of the system is re-absorbed into the blood stream and a concentrated urine is formed. Conversely, when a lot of water is drunk the osmotic pressure of the blood falls, the hormone is not released and the walls of this part of the tubule and duct become impermeable to water, so much larger amounts reach the bladder and the urine is dilute. In this way the kidney acts as an osmo-regulatory organ.

The amount of urea in the blood is also kept fairly constant by the kidneys. When we eat a large amount of protein the concentration of urea in the blood rises due to deamination. (Would you expect any significant differences in urea levels between a child and an adult?) The kidney adjusts by passing more urea into the urine. However, this is not a precise mechanism, as not all the urea passing into the glomerular filtrate finds its way into the urine; some is re-absorbed. The main point is that the kidneys reduce blood urea to a level where it is not harmful.

Maintaining the level of blood sugar

Sugar, being an osmotically active substance, has to be carefully controlled, otherwise irreparable damage would be done to the brain and body tissues.

It is significant that the sugar absorbed from the gut is first taken to the liver and, if not immediately required, converted to glycogen as osmotically inert, insoluble granules. Although the input of sugar is irregular, the body requires a continuous supply for respiration. One of the functions of the liver is to ration out the sugar according to the body's needs. This involves the reconversion of glycogen to glucose. The process is reversible and is controlled by several hormones, one of which is **insulin**. Insulin is formed by the pancreas in special areas known as **islets of Langerhans**. People who are unable to produce enough insulin suffer from **diabetes mellitus**. With this condition the blood sugar level rises above normal and as a result some sugar appears in the urine. Thus diabetes can be diagnosed by testing the urine for sugar with Benedict's solution.

Diabetes is a relatively common condition, but it varies in its severity. The normal treatment is to have regular injections of insulin prepared from the pancreas of animals. The quantity needed has to be accurately calculated according to the amount of excess sugar in the blood, and the diet is regulated to keep the sugar intake as constant as possible and at a low level.

Homoiothermy

This is the maintenance of a fairly constant temperature irrespective of the environment. Only the birds and mammals can do this efficiently and so they are called **homoiothermic** or warm-blooded in consequence. 'Warm-blooded' is not a very good term as a so-called 'cold-blooded' animal basking in the sun on a hot stone may have a higher temperature than a warm-blooded animal. It is the maintenance of a constant temperature that is the important point. Homoiothermy brings many advantages to birds and mammals. It enables them to live in places where temperature conditions are extreme, such as in polar regions and hot deserts. It also ensures that their metabolic processes are carried out at a steady and rapid rate, allowing them to be active under conditions where cold-blooded animals would be torpid.

Different species keep their blood temperatures at different levels. Our so-called 'normal' temperature is 36·9°C (98·4°F), a hen's is

39·4°C (103°F), a blue-tit's 41·7°C (107°F), and a humming-bird's 43·3°C (110°F). Consider what differences these temperatures would make on their metabolic rate and what consequences this could have on their activities.

When we say our normal temperature is 36·9°C (98·4°F) we mean this is the *average* figure for the human population as a whole.

Find out, through a class experiment, how much the 'normal' varies from person to person. Clinical thermometers used by doctors are the best, but if there are not enough to go round they should be washed in disinfectant before being used again. From the class results make a bar graph (histogram) recording how many in the class have temperatures in each 0·2°F range. How does this graph compare with the population average? If it varies, can you suggest why?

You would also find it interesting to study the fluctuations in your own temperature by taking it at two-hour intervals throughout the day and plotting the results on a graph. You should start as soon as you wake up and note on the graph your activities just before you take each reading.

The temperatures of hibernating mammals such as bats, dormice and hedgehogs, fluctuate much more than ours. In winter they become almost cold-blooded, their body temperature dropping to a degree or so above their surroundings, but in summer their temperature is well above this and fluctuates very little. However, bats are peculiar because in daytime, when they go back to their roosts (caves, church towers, etc.), their temperature quickly

Fig. 6:5 A hibernating dormouse. Why is its curled-up condition significant?

drops; this means they have a daily cycle of temperature change as well as an annual one. Why is it advantageous for small, active creatures like bats to become cold when they are inactive, and how could this be disadvantageous?

Heat gain and loss

In order to maintain a constant temperature, the heat lost from the body to the external environment must be replaced by heat generated within the body. We have already seen that energy in the form of heat is released when food is broken down during respiration. It is this heat that warms our bodies. Some parts, such as the muscles and liver, release more heat than others, but even distribution is effected by the blood which transports the heat all over the body. Obviously more will be needed by those parts which are losing heat most quickly.

The body is continually losing heat: through the skin by radiation or conduction or by sweating, and to a lesser extent through breathing, urination and defaecation.

It is the skin which provides the greatest surface area of the body which is in contact with the environment, so we should expect to find the most important heat-regulating mechanisms there.

Structure of the skin

You will see from Fig. 6:6 that the skin consists primarily of two layers, the **epidermis** and **dermis**. The epidermis also consists of several layers: 1. The **Malpighian layer** composed of actively dividing cells which are constantly replacing the rest of the epidermis. 2. The **granular layer**, consisting of living cells which are the products of the Malpighian layer. 3. The **cornified layer** which is constantly being formed from below when the granular cells become horny and die. This dead layer is tough and protects the living cells from mechanical injury and fungal and bacterial attack. It also reduces the amount of water lost from the surface. On the soles of the feet and on the palms of the hands this layer may become very thick, especially in people who walk bare-footed or use their hands for heavy work, or for a particular activity such as guitar-playing. The outer portion of this layer is

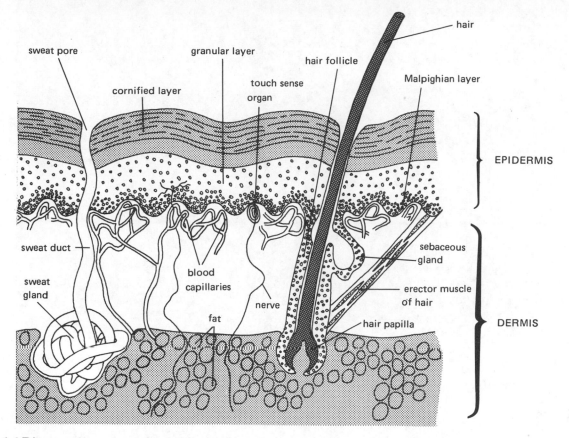

Fig. 6:6 Diagram of a section through human skin.

constantly being worn away; washing removes some of it, and the scurf on our heads which we may notice when we brush our hair is a sign that this is happening.

The **hair follicles** are pits lined by epidermal cells. The cells at the base, or **papilla**, are constantly multiplying, forming hair cells within the follicle. The constant adding of cells to the base of the hair causes it to grow, but the hair cells quickly become horny and die so that it is only the base of the hair that is living.

Opening into the follicles are **sebaceous glands** which secrete an oily substance which keeps the skin and hairs supple. Vigorous brushing of the hair stimulates the glands and this improves the condition of the hair—combing is no substitute. The fluid produced, known as **sebum**, also has antiseptic properties and in aquatic mammals especially it is very useful as a water repellant. For example, when an otter comes out of the water and shakes itself, its fur becomes dry very rapidly.

The **sweat glands** are present all over the body, but are especially numerous in the palms of the hands, the soles of the feet, under the arm pits and between the thighs. Each gland is a much coiled tube which lies deep in the dermis; it is connected to the surface by a spiral duct which opens on to the surface as a sweat pore. The cells of the sweat gland receive fluid from the surrounding blood capillaries and secrete it into the duct through which it passes to the surface of the skin. Sweat is a fluid which can be looked upon both as a secretion and an excretion; a secretion because it serves a useful purpose when it evaporates and cools the body; an excretion because it also contains some urea which is a harmful metabolic product. Sweat also contains some sodium chloride which is useful to the body; so people who sweat a great deal because of their occupations, or because they live in the tropics, need to take extra salt, either as tablets or with their food, to keep the salt and water balance of their body fluids correct.

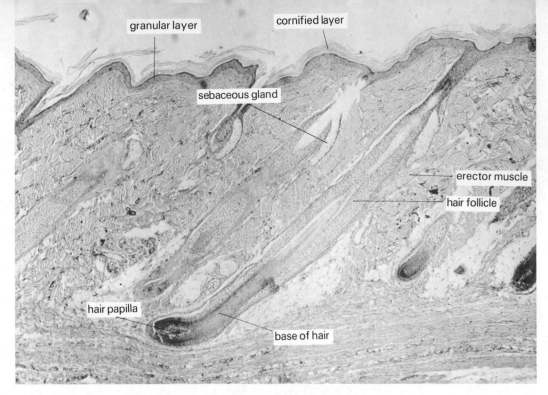

Fig. 6:7 Photomicrograph of a section through human scalp.

The **dermis**, which is the inner layer of the skin, is made up of connective tissue and contains many blood capillaries, lymph vessels, sense organs which detect touch stimuli and temperature changes (p. 102), and elastic fibres which bring the skin back to shape when it is stretched or distorted. In the outer layers of the dermis there are cells containing the pigment **melanin** which protects the under-lying tissues from the damaging effect of ultra violet light. When fair-skinned people become tanned by the sun it is due to an increase in melanin. Dark-skinned people who live in tropical countries have large quantities of melanin in their skins, a characteristic which is inherited. There are also muscles attached to the hair follicles. When contracted these muscles erect the hair and cause the skin to be covered in 'goose pimples'. Below the dermis there is usually a layer of stored fat, which acts as an insulator as well as a food store. In whales this layer is incredibly thick and constitutes the blubber from which whale oil is made.

Temperature regulation

First let us consider how the skin reacts when the body becomes too hot. In the brain there are temperature-sensitive cells which act like a thermostat. If the blood becomes hotter than the 'normal' temperature, on reaching the brain it stimulates certain cells and nerve impulses are sent to the skin with two different results: 1. The walls of the arterioles in the dermis increase in diameter—this is called **vaso-dilation**. This allows more blood to flow into the capillaries near the surface of the skin so more heat is lost from the body by radiation. This is the reason why we go red when we are hot. 2. The rate of sweat production is in-creased and as the sweat evaporates it takes heat from the body (latent heat of vaporisation).

What happens when we become too cold? Our reaction is to shiver; this liberates heat when the muscles contract. Sometimes we stamp up and down and do some exercise; this has the same effect as shivering. But apart from the generation of more heat, the body responds by decreasing heat loss. There are two important methods: 1. The walls of some of the arterioles in the dermis contract, a process known as **vaso-constriction**. This has the effect of reducing the blood supply to the capillaries near the surface, so far less heat is lost through radiation, and as a result we go pale and sometimes blue with cold. 2. Sweat production is reduced to negligible amounts, so less heat is lost in consequence.

Temperature reducing mechanisms are more effective if the surface area exposed to the atmosphere is increased and if the air is moving over the surface (or the surface itself moves). This is why African elephants flap their huge ears to help cool themselves; by so doing they increase their body surface by a third! You will have seen dogs panting with their tongues hanging out when they are hot. There are no sweat glands in the tongue; how do you think this helps them to cool down?

An interesting case of temperature regulation occurs in whales. The thickness of the blubber varies in different species, the more active ones having thinner layers of blubber. This is because the blubber is such a highly efficient insulator that during bursts of great activity so much heat is generated internally that the whale is in danger of cooking inside its jacket of fat unless the heat can be dissipated in some way. The problem is solved by the presence of a complex system of blood capillaries within the blubber. These dilate when the temperature of the blubber increases, so allowing a greater volume of blood to pass through and increasing the rate of heat dispersal. When a whale dies this mechanism cannot operate and after a time decomposition by bacteria raises the temperature of the body so much that when the whale is cut open the tissues are found to be blackened by the heat!

It follows from what we have learnt about the principles of maintaining a constant temperature that if it is necessary to conserve body heat because the habitat is cold, it is better for warm-blooded animals to have as little surface as possible compared with their volume. In other words, large animals would be more successful than small ones in a cold climate. Find out if this is true by making a list of the Arctic and Antarctic birds and mammals and see if you can find many small ones.

Another aspect of this is the fact that the smaller animals have a relatively large surface area and so they must eat a lot of energy food to make up for the heat loss. Consequently small animals such as shrews have to eat more than their own body-weight of food every day!

We referred briefly to insulation as a means of keeping in heat and mentioned the fat layer under the skin. In addition, the air around the skin is an excellent insulator if it is not moving. Fur and feathers serve this purpose admirably as they trap air close to the skin, and when a mammal or bird fluffs up its coat or feathers more air is trapped and the insulation is made more effective.

If you have understood these principles you should be able to work out the reasons behind the following facts:

1. Hedgehogs and dormice curl up when they hibernate, and you do the same when you get into a cold bed.
2. Eskimos are short, broad and fat people.
3. The Arctic hare and Arctic fox have much smaller ears than the brown hare and the common fox which occur further south.
4. A string vest under a cotton shirt keeps you warm in spite of the large holes in the vest.
5. A wind-cheater is used by mountain climbers to prevent sudden chilling.
6. Two thin pairs of socks are often warmer than one pair of thick ones.
7. Nylon shirts make you feel very hot in tropical countries, while cotton shirts are cooler.

To summarise: the skin is the main barrier between an organism and its environment. Its main functions in mammals are a) to achieve homoiothermy, b) to protect the underlying tissues from mechanical injury, too much water loss, the effect of ultra violet light and invasion by bacteria and fungi, c) to act as an accessory excretory organ for eliminating some urea.

There are, of course, additional functions of the skin which are important. Vitamin D can be synthesised in the skin of mammals through the effect of ultra violet light (Book 1 p. 164). The skin also makes an organism aware of its surroundings through its sense organs which detect touch, pain and temperature change. This function will be discussed in more detail in Chapter 8.

7

Support and movement

Support systems

All organisms need some support for their bodies and the larger they grow the greater this need becomes.

The method developed by most plants (Book 1 p. 124) is through having rather rigid cell walls made of cellulose; the turgidity of these cells helps to give rigidity to the whole plant. This is aided by the presence of a tough cuticle on the outside and bundles of tough woody fibres and vessels inside. When woody elements are formed in dense masses, as in shrubs and trees, they play a major part in providing support. Although extremely effective as supporting structures, their rigidity prevents all forms of locomotion; however, plants have no need to move as they can feed without doing so.

Animals, by contrast, usually have to move to find their food and they have evolved methods of support which still allow freedom of movement. Most simple animals have little need for a support system as the majority are very small and many are aquatic, so the water buoys them up. Even some of the larger ones such as jelly fish can obtain enough support from the water to allow them to float. Do you think their shape would be significant in this respect?

Terrestrial animals such as earthworms gain support from the fluid inside the coelom. This fluid exerts pressure on the body wall, and so makes it fairly rigid. As a result the animal is able to move by means of rhythmic contractions of the body wall using the fluid

'skeleton' as a support (Book 1 p. 40). Most land animals have evolved a more rigid system which not only gives much greater support but aids locomotion as well. The arthropods use an exoskeleton (Book 1 p. 56) but this has a major disadvantage in that when the animal grows it has to split this hard covering and grow another.

The vertebrates, on the other hand, have evolved an endoskeleton composed either of cartilage or cartilage and bone, which has the advantage of being able to grow internally without interruption.

THE MAMMALIAN SKELETON

The skeleton has three main functions: 1. To give support to the rest of the body and enable it to retain its shape. 2. To protect vital and easily damaged parts of the body from injury. 3. To help in locomotion—first, by having joints which give flexibility to the body, second, by providing a firm foundation for the attachment of muscles.

Before studying in more detail how the skeleton fulfils these functions we will first consider the tissues of which the skeleton is composed.

Cartilage and bone

Cartilage and bone, like all tissues, are composed of living cells and a non-living secretion or **matrix** which is produced by these cells. The matrix makes these tissues tough. Cartilage is more flexible than bone as the matrix is made of protein; bone also has a protein matrix, but in it calcium salts are deposited which give it much greater rigidity. You can see for yourself how efficient these salts are in giving rigidity to a bone in this way:

> Obtain a rib bone from the butcher and clean off any meat. Place it in a glass cylinder containing dilute hydrochloric acid and leave it for several weeks. The acid will gradually dissolve out the calcium salts but will not affect the protein matrix. Finally, wash the bone thoroughly and test its rigidity.

You will see from Fig. 7:1 that in bone tissue the cells are arranged in concentric circles around blood vessels, and the bony

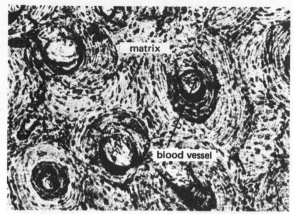

Fig. 7:1 Photomicrograph of a thin section of compact bone. The black dots arranged concentrically are the spaces in which the bone cells lie.

volving a gradual removal of cartilage by cells from outside which invade it; other cells of a different kind then follow and lay down bone which replaces the cartilage which has been removed.

If you examine an X-ray of the limb bones of a child (Fig. 7:2) you will see a region between the head and shaft of each bone where cartilage is still present. These are the places where growth in length is still taking place. It is possible to estimate the age of the child from the size of these regions.

If a child's bone is fractured, it may bend and split on one side only, instead of breaking completely, because it is incompletely ossified; this is known as a **greenstick fracture**.

matrix laid down by these cells is in the form of a series of long cylinders. Cartilage, by contrast, contains no blood vessels within its matrix, although they are present in the surrounding sheath of tissue, so some diffusion of oxygen and nutrients can take place between the blood and the cartilage cells.

In the foetus the skeleton is first laid down as cartilage, but as development proceeds it is replaced by bone—a process known as **ossification**. This is a complex process in-

Fig. 7:2 X-ray photograph of the knee joint of a child aged 8. Note the growth regions where cartilage is still present.

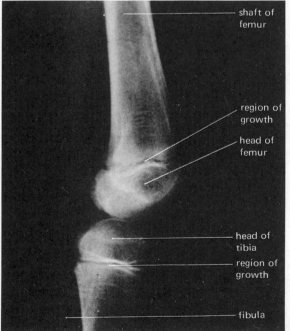

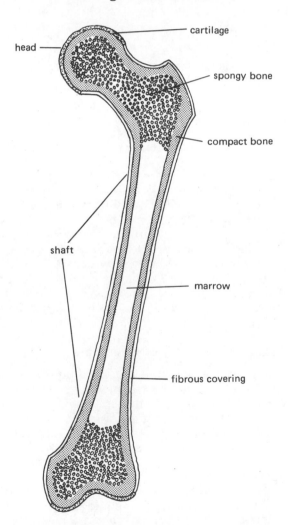

Fig. 7:3 Longitudinal section of a femur (diagrammatic).

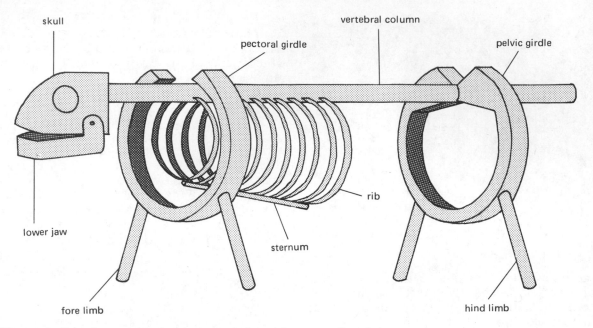

Fig. 7:4 Stylised diagram showing the general plan of the mammalian skeleton.

Structure of a bone

Bones vary in shape, size and strength according to the functions they serve. A long bone, e.g. the femur (Fig. 7:3), has to be strong to give support to the body, but light enough to allow proper movement. This is achieved by having the hardest area of bone, called **compact** bone, on the outside to form a rigid cylinder, with looser tissue known as **spongy** bone further in. The hollow centre is filled with marrow where blood cells are manufactured and fat is stored. The ends are particularly strong and solid and are adapted to articulate with another bone at the joint (p. 86).

The general plan of the skeleton

All vertebrate skeletons are based on the same general plan. The skeletal plan for a mammal, when reduced to its simplest form, is shown in Fig. 7:4. Basically it consists of:
1. An **axial** part which runs down the length of the body; this consists of the skull, the vertebral column with its ribs, and the sternum or breast bone.
2. An **appendicular** part which consists of the girdles and limbs. The shoulder or **pectoral** girdle and the hip or **pelvic** girdle articulate with the fore and hind limbs respectively.

Carefully compare this simple diagram with that of the human skeleton (Fig. 7:5). The latter will give you details of the arrangement and the names of the major bones.

Examining the skeleton

It is not easy to obtain a human skeleton, although plastic models are sometimes available, but you should examine a mounted rabbit's skeleton to see the shape and position of the bones and try to deduce the main functions which they serve.

Also, if at all possible, examine and mount on card the bones of some of the small mammals. As explained in Book 1, these may be obtained either from owl pellets or from old milk bottles discarded at lay-bys etc. Small mammals enter these bottles through curiosity and fail to get out again, so complete skeletons may be obtained in this way. Wash the bones thoroughly, bleach by leaving them over-night in a dilute solution of hydrogen peroxide and then stick them on to cards arranged, as far as you can, in their correct positions. Although they are very small, you should be able to recognise the bones by comparing them with those of a mounted rabbit's skeleton.

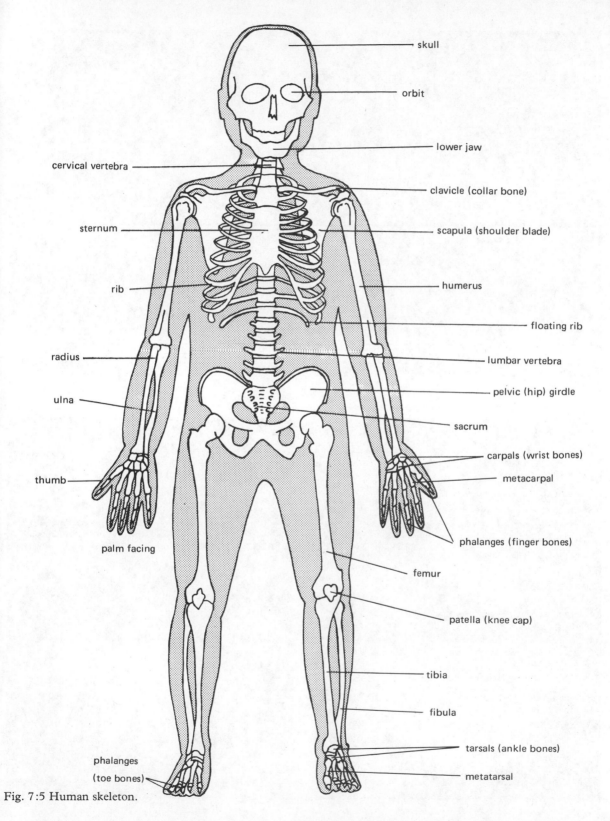

skull

orbit

lower jaw

cervical vertebra

clavicle (collar bone)

sternum

scapula (shoulder blade)

rib

humerus

floating rib

radius

lumbar vertebra

ulna

pelvic (hip) girdle

sacrum

carpals (wrist bones)

metacarpal

thumb

phalanges (finger bones)

palm facing

femur

patella (knee cap)

tibia

fibula

tarsals (ankle bones)

phalanges
(toe bones)

metatarsal

Fig. 7:5 Human skeleton.

83

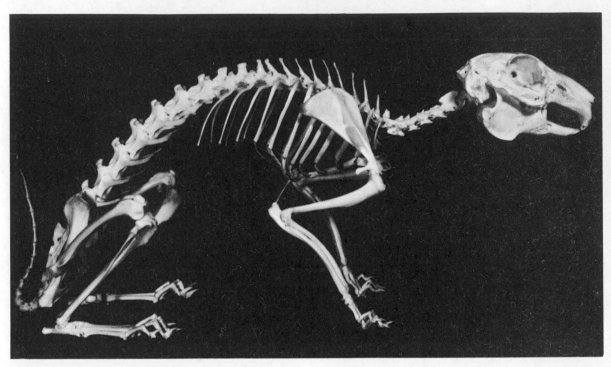

Fig. 7:6 (above) Rabbit skeleton.
Fig. 7:7 (below) Human lumbar vertebra, end-on.

Fig. 7:8 Human lumbar vertebra, side view.

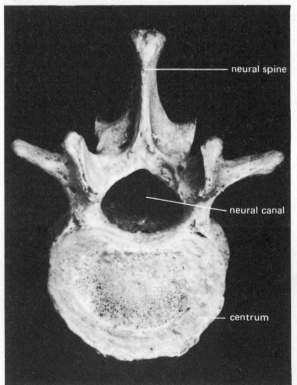

neural spine

neural canal

centrum

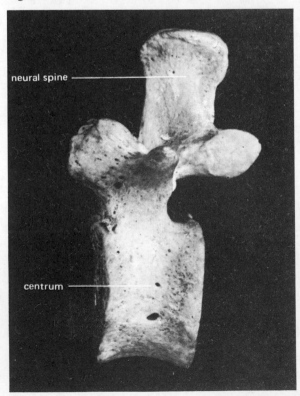

neural spine

centrum

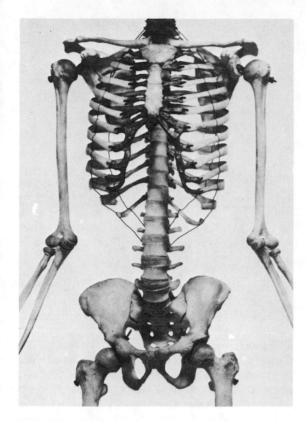

Fig. 7:9 Human rib cage and pelvic girdle.

The protective function of the skeleton

The skeleton, along with its attached muscles and ligaments, provides wonderful protection for the vital organs. The casualty wards of hospitals are full of people who are fortunate to be alive after serious road accidents, thanks to this protection, although there are limits to the shocks that bones can stand.

The skull is really a compact group of boxes fused together to form a single unit with the jaws attached. The largest box is the **cranium** which protects the brain and the smaller ones clustered round it are the **sense capsules** which protect the main sense organs—ears, nose and eyes. For obvious reasons, the eyes cannot be completely enclosed and so they are more vulnerable in consequence.

Similarly, the delicate spinal cord which leads from the brain is protected by the **vertebral column** which consists of a series of strong rings of bone firmly attached to each other by ligaments. Between the brain and spinal cord and their protective bony coverings are membranes enclosing a jacket of fluid —the **cerebro-spinal fluid**—which bathes these delicate organs and acts as a shock absorber. Compare this with the amniotic fluid which protects the foetus (p. 31).

The heart and lungs are well protected by a cage-like structure composed of the vertebral column, ribs and sternum, together with their attached muscles. The floor of this cage is formed by the diaphragm which separates the contents of the thorax from the abdominal organs. The thoracic cage must be strong, but it must also allow movement for breathing (Book 1 p. 137). This flexibility is aided by the ribs having a cartilaginous portion where they meet the sternum.

The pelvic girdle also supports and protects the organs in the lower part of the abdomen; this is particularly important during pregnancy.

MOVEMENT

We saw, when studying insects, that a rigid exoskeleton made locomotion and movement of the parts impossible without the presence of joints. The same applies to our own endoskeleton. We have about 200 separate bones in our body and these meet each other at joints.

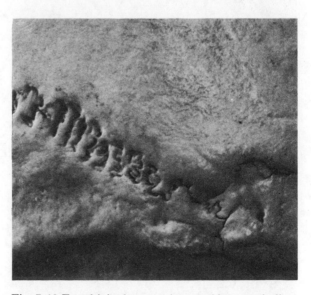

Fig. 7:10 Fused joint between bones of human skull.

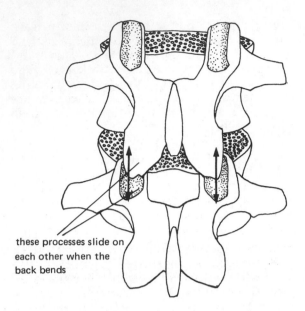

these processes slide on each other when the back bends

Fig. 7:11 (above) Sliding joint between lumbar vertebrae, dorsal view.
Fig. 7:12 (below) Ball and socket joint of the hip.

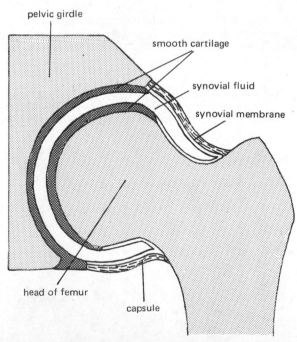

pelvic girdle

smooth cartilage

synovial fluid

synovial membrane

head of femur

capsule

There are two main kinds of joints:
1. **Fused joints** where the bones are rigidly fixed to each other by a pad of fibrous tissue, as in the bones of the skull (Fig. 7:10).
2. **Synovial joints** where friction between the bone surfaces is greatly reduced by the presence of a lubricating **synovial fluid**. These can be classified further according to the degree of movement the joint allows:

a) **Sliding joints** where only a very limited amount of movement is possible, as between the joints of the wrist and ankle (Fig. 7:11).

b) **Hinge joints** which allow considerable movement, but in two planes only.

c) **Ball and socket joints** which allow most movement of all, in three planes.

Test out the joints in various parts of your body to see which categories they come under. Include the shoulder, elbow, base of thumb, finger, hip and knee. In some cases it is best to hold one part still while you move the other. List your results.

A ball and socket joint is a beautiful piece of engineering (Fig. 7:12). At a butcher's you will often see the ball of a joint exposed as a glistening white knob. This covering is composed of an extremely smooth cartilage and it fits into the socket which has a similar surface. The **synovial fluid** which lubricates the junction between these surfaces is secreted by the **synovial membranes**. The articulating bones are held in position by strong ligaments which are sufficiently elastic to allow just enough adjustment when the bones move. There is also a capsular **ligament** which acts like a sleeve round the joint; this keeps the bones firmly in place and helps to contain the synovial fluid.

If you **dislocate** your shoulder or finger the bones are displaced at the joint. Dislocations stretch the ligaments abnormally and may damage them and as a result the joints are slightly looser afterwards. Unfortunately, this makes further dislocation of the same joint more likely. If this happens repeatedly it is sometimes possible for a surgeon to 'tighten' them.

When you jump from a height, on landing the main joints concerned have to withstand a considerable shock. The danger of damage is reduced in two ways. Cartilage is not so rigid as bone, having more resilience, so that the cartilage of the articulating surfaces also acts as a shock absorber. Secondly, between the vertebrae there are discs of tough fibrous cartilage which act as cushions. In the centre of these discs is a bag of fluid. If the vertebral

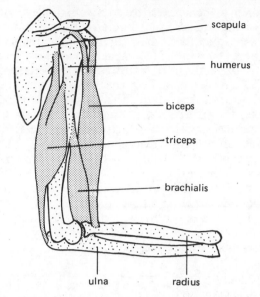

Fig. 7:13 The elbow joint showing the positions of the antagonistic muscles.

column is given a violent twisting movement, this bag of fluid may be squeezed sideways as a projecting bump, causing great pain if it presses on a nerve. This condition is known as a slipped disc.

Muscles

Bones do not move by themselves; the contractions of muscles cause them to move. It is because muscles only do work when they contract that at every joint there has to be at least one muscle which moves the bone in one direction and at least one other which brings it back again. Muscles which oppose each other's action are said to be **antagonistic**. When you move your arm (Fig. 7:13), there are two main muscles, the **brachialis** and the **biceps**, which bend the arm (for this reason they are described as **flexor** muscles), and a **triceps** muscle at the back which straightens it again (this is called an **extensor** muscle). Most joints in the body have at least one flexor and one extensor muscle. When the flexor muscle contracts the extensor is in a relaxed state, and vice versa.

Test this action for yourself by holding your right arm straight with the palm upwards. Grasp the upper arm with your left hand so that your fingers touch the biceps from above and your thumb touches the triceps from below. Now slowly raise your arm. What is happening to the muscles above? Now straighten your arm. What happens to the muscle below?

You will have noticed that when muscles contract they also become fatter. Each muscle (Fig. 7:14) is composed of many bundles of elongated fibres bound together by connective tissue which also forms a sheath round the whole muscle. The sheath is extended at each end into a **tendon** which attaches the muscle to the bone. When a muscle contracts, its individual fibres become shorter and fatter due to the special properties of two proteins which they contain called **actin** and **myosin**.

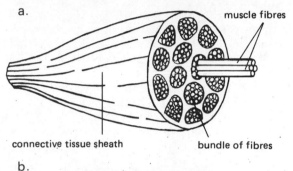

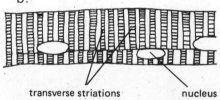

Fig. 7:14 The structure of a muscle: a) part of a muscle b) part of two muscle fibres much enlarged.

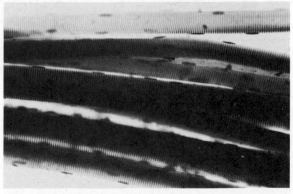

Fig. 7:15 High power photomicrograph of muscle fibres separated out.

87

The stimulus which causes them to contract comes from the central nervous system (p. 104) and the energy is provided when food is broken down in respiration. Consequently, muscle must have a very good blood supply to bring the sugar and oxygen needed for respiration and to take away the waste products. Under the microscope you can see the blood capillaries supplying the various fibres. Much can be learnt about the structure of muscle by observing meat obtained from the butcher, as most meat is animal muscle. A good carver will cut red meat so that the small bundles of muscle fibres are sliced transversely; the connective tissue between them allows thin slices to be made.

Lever systems

The movements caused by muscles at a joint can be explained in terms of lever systems. There are three kinds of levers according to the position of the pivot or fulcrum and the place where the force is applied. All three types of levers are used in the body. Figure 7:16 gives an example of each.

For a lever system to work efficiently one end of the muscle has to be attached to something rigid so that when the muscle contracts the bone to which it is attached at the other end is the only part to move. The fixed end of the muscle is called the **origin** and the other end is known as the **insertion**. You will see that the biceps muscle has its origin in two tendons which are attached to the firm scapula and its insertion is on the radius. The brachialis has its origin on the humerus and its insertion on the ulna.

Some forms of lever have a greater mechanical advantage than others in that less force has to be exerted to move a particular weight. You can work this out for yourself by constructing a simple model of your elbow joint.

Set up the apparatus as in Fig. 7:17. The ruler represents the arm which is pivoted at the elbow and carries a weight in the hand. The spring balance takes the place of the muscle and the force needed to keep the arm in a horizontal position can be read off. Record what force is needed to keep the arm horizontal

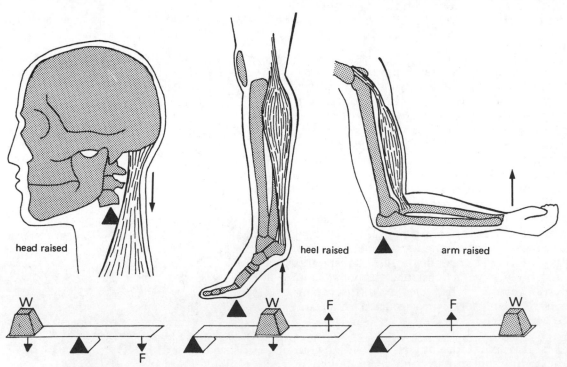

head raised   heel raised   arm raised

Fig. 7:16 Examples of lever systems in man. In each case the contraction of the muscle provides the force and moves the bone, and the joint acts as a fulcrum. W = weight, F = force, ▲ = fulcrum.

88

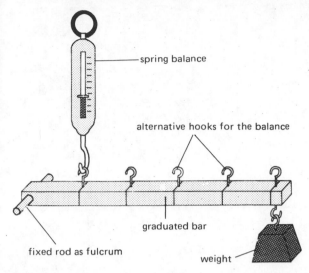

spring balance

alternative hooks for the balance

graduated bar

fixed rod as fulcrum

weight

Fig. 7:17 A model which simulates the lever action of the forearm.

when the spring balance is attached at different equally-spaced positions along the ruler. Repeat with different weights.

In the light of your results consider Fig. 7:13 again, noting the points of attachment of the muscles. Would another arrangement have given a greater mechanical advantage? If so, why do you think the attachment is where it is? One clue is to consider the distance the hand would move for different points of insertion, assuming that the brachialis and biceps muscles contract the same amount each time. Is greater mobility attained by having the point of insertion near the elbow joint?

Tendons

These connections between muscles and bones are composed of very fine fibres which are extremely strong and do not stretch. In the butcher's you see them as shiny white sheets or cords attached to the meat (muscle). You can demonstrate for yourself the action of tendons and their properties in this way:

Examine a fresh chicken's foot. You will see one of the large tendons projecting from the cut end. What happens when you pull this tendon? Follow the tendon down to the toes by cutting away the skin carefully with a scalpel. Can you now explain the movements you previously observed?

The attachment of a tendon to a bone is more effective when the bone surface has grooves and projections. This is the reason why bones to which the tendons of large muscles are attached have such strange shapes. Note all the projections from the various kinds of vertebrae. The lumbar vertebrae have especially large ones which aid the attachment of the powerful back muscles (Fig. 7:7). The relationship between bone shape and muscle is so precise that the shapes of extinct animals known only from their bones (e.g. dinosaurs) can be deduced by the position of muscles as indicated from projections on fossil bones.

Locomotion

We have now seen how bones and muscles with their various joints and lever systems function in moving parts of the body. Let us consider further how this leads to locomotion of the whole organism.

In vertebrates, movement results from the backward thrust of the limbs against the ground, water or air according to where the animal is living. In order to move the whole body this force has to be transmitted to the spine. This is done through the two girdles. As we walk and run in an upright position it is only the pelvic girdle which plays a major part. This is fused to the sacral region of the vertebral column by very strong joints (Fig. 7:9).

The pectoral girdle is *not* fused to the spine, but instead there are strong muscles which attach it to the thoracic vertebrae. It is thus less effective in transmitting the force from the limbs to the body. It is therefore not surprising that in animals which run on all four feet the hind limbs produce the greatest thrust. The pectoral girdle with its muscular attachment to the back bone provides an excellent shock absorber in these animals when they land after a leap.

There are many factors which influence the effectiveness of the thrust and hence the speed of the animal or its jumping ability. Consider some of the mammals which are noted for speed, such as the cheetah or greyhound, and others for their jumping ability such as the kangaroo, rabbit and gazelle. Which of the following factors do you consider are important in each case:

1. The weight of the animal?
2. The distribution of the weight?
3. The length of the bones?
4. The number of joints, i.e. levers?
   Remember that muscle itself is heavy.

Check your conclusions by considering the build of Olympic sprinters and high jumpers. Is there any skeletal reason why coloured athletes are so often the world's greatest sprinters?

Posture

Muscles are important not only for movement but also for maintaining posture. You can test this for yourself.

Stand upright with your hands at your sides and close your eyes. As you consciously keep yourself upright you will feel the muscles in various parts of your legs contracting to keep you in position. Usually this happens quite unconsciously. Figure 7:18 shows how the main muscles are used for this purpose: notice how they support the main joints concerned—the hips, knees and ankles.

Balancing the head on the top of the spine (when we stand upright) is helped by the curvature of the spine. The head is very heavy and its position is controlled by the large neck muscles. When we sit at a desk we sometimes hunch ourselves up; this puts a lot of strain on these muscles and we often compensate by propping our head up with our arm. If our posture is right this should not be necessary. Developing a good posture early in life and exercising regularly the muscles concerned is the best insurance against 'aches and pains' in older age.

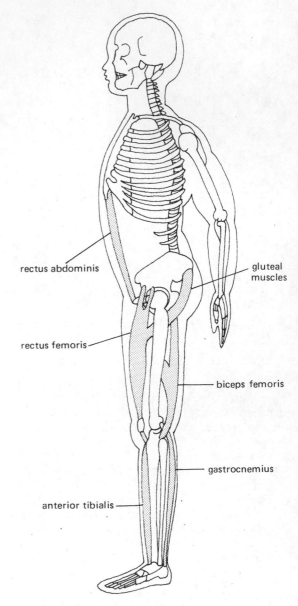

rectus abdominis

gluteal muscles

rectus femoris

biceps femoris

gastrocnemius

anterior tibialis

Fig. 7:18 Diagram showing the main muscles concerned in keeping an upright posture.

8

Sense organs

Responding to the environment

An organism cannot exist in isolation; all the time it is subjected to many influences which come from its surroundings. These influences constitute the **environment** in which it lives. Many of these factors are physical, such as temperature, light, humidity and the nature of the soil, but equally important are the activities of all the other living things around it. As we saw in Book 1 all organisms, wherever they live, are adapted in their structure and behaviour to the particular conditions of their environment. Most people live in towns and cities, an environment of steel and concrete, parks and gardens, cats and starlings, cars and buses, and above all, lots of other people. All these factors which influence us are changing constantly and we are responding all the time in one way or another to these changes. We have, in addition, an internal environment which is also changing—the composition and temperature of our blood, the quantities of hormones being secreted by our glands, the activity of our cells and so on. We are constantly adjusting to these changes taking place inside us, and we discussed some of the mechanisms involved in Chapter 6.

All these changes, both external and internal, are known as **stimuli**. For an organism to respond to these stimuli there are three requirements:

1. A means of detecting the changes. This is the function of the **receptors** or **sense organs**. They are the receivers of information from the environment. The main sense organs are large aggregations of cells which are sensitive to particular stimuli and include the eyes, ears, nose and tongue.

2. A means of acting on the information received. This is the function of the **effector** organs such as muscles which bring about movement and glands which secrete fluids.

3. A linking system within the body which ensures that the stimulus received is followed by an appropriate response. This is the function of both the **nervous** and **endocrine** systems.

In this and the following chapter we shall be studying these aspects in more detail, after which we will proceed to a study of **behaviour** —the response of an organism to all the information it receives from its environment, both internal and external.

SENSE ORGANS

What sort of stimuli do we receive?

Everything we know about the world we live in results from the information we receive from our sense organs. It is too simple to say, as **Aristotle** did, that we have just five senses— sight, hearing, smell, taste and touch—for the biologist recognises more than these. For example, a blind person when he touches an object not only discovers that it is there, but he can also detect its size and shape, its texture and its temperature, and if he touches a liquid, he knows whether it is slippery or viscous. He can distinguish between touching and being touched and between different intensities and pressures.

Consider your other sense organs—your eyes and ears, for example. What information do they provide apart from detecting light and sound?

The more information we receive from our environment the greater the possibilities of our making the most appropriate responses. We receive no information from our sense organs of touch or pressure, taste or pain until contact with the object is made. But with sight, hearing and smell it is quite different as these stimuli come to us from a distance—sight and sound very rapidly, smell more slowly. Thus some sense organs extend our knowledge of the environment far beyond the limits of our immediate surroundings.

We can distinguish at least six kinds of sense organs according to the type of stimuli they receive:

Stimulus	Receptor	Function
1. Light	Eye	Seeing
2. Sound	Ear	Hearing
3. Chemicals	Nose and tongue	Smelling and tasting
4. Temperature	Skin	Detecting temperature (both hot and cold)
5. Mechanical stimuli	Skin and muscles	Feeling and gauging pressures
6. Gravity	Ear	Balancing

Another way of classifying sense organs is to divide them into:

1. **Exteroceptors** which receive the stimuli from outside the organism, e.g. eyes, ears, touch organs and taste buds.

2. **Enteroceptors** which receive stimuli from inside the body such as those which detect changes in the composition of the blood.

3. **Proprioceptors** which are present inside muscles and tendons and act like pressure gauges; these are concerned with posture (p. 90).

We will now consider the main sense organs in turn, concentrating especially on the eyes and ears.

THE EYE

Kinds of eyes

Light receptors in different animals range from the light-sensitive areas in the cytoplasm of *Euglena* and the scattered cells in the skin of the earthworm, to the complicated compound eyes of insects and the highly efficient camera-like eyes of vertebrates. The simplest structures can only detect variations in light intensity, but the highest forms produce images which provide information about the size, shape and even the colour of the object, as well as its distance from the observer. Let us now study our own eyes.

How do we protect our eyes?

Feel the bone all round the eye; this forms the rim of the eye socket or **orbit** in which the eye is sunk. A boxer would quickly be blinded without this protection.

The eyelids also protect the eyes from foreign particles; in a dust storm one would almost close the eyelids to add to their efficiency. In addition, the eyeball is covered externally by a very thin transparent membrane, the **conjunctiva**. If dust gets into the eye this membrane becomes inflamed and pink.

The eyebrows and lashes act like the lens hood on a camera, protecting the eye from glare. They also help to prevent rain or sweat from entering the eye.

Blinking not only protects the eye from mechanical injury, but is protective in another way. Every time you blink, fluid from the tear gland under the upper eyelid is spread over the eye surface, excess draining away through the tear duct to the back of the nose (Fig. 8:1). Ask your neighbour to pull down the lower lid and examine the corner of the eye nearest the nose. You will see the opening of the tear duct and also the pink mass which represents the remains of a third eyelid (it is functional in birds). Tears not only wash away dust but being antiseptic they help to prevent bacterial infection of the eye.

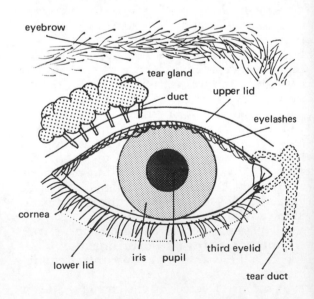

Fig. 8:1 The human eye showing the position of the tear gland and ducts.

External features of the eye

Look your partner in the eye. The white part is the **sclerotic**, a tough protective membrane which surrounds the eye. In front it becomes transparent to let the light through; this part is called the **cornea**. You will not see the conjunctiva which covers the cornea as it is transparent. The coloured part is the **iris**, a sheet of muscle which controls the size of the hole in the middle which is the **pupil**.

The iris is composed of muscle fibres which are arranged both radially and in a circular manner (Fig. 8:2). When the circular muscles contract the pupil becomes smaller; when the radial muscles contract it becomes dilated. In bright light the pupil is reduced to a tiny aperture, so protecting the retina from damage, while at night the pupils are very dilated allow-

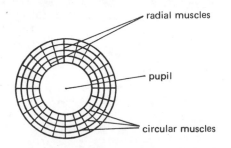

Fig. 8:2 The iris seen from in front showing diagrammatically the arrangement of muscles.

ing as much light as possible to enter. The larger the eye, the more the pupil can dilate and the more light is able to enter. This is why nocturnal animals usually have large eyes. Test the reaction of your own eyes to light in this way:

This experiment works best in a dimly-lit room. First, notice the size of your partner's pupils, then flash a torch near the eye and note any change. Now hold a book between the eyes to shade one eye while you flash the light on the other. What happens to the pupil of the shaded eye? What do you conclude from this?

Eye movements

Try these movements:

Stand in front of your partner who must keep his head quite still, and ask him to follow various movements of your hand with his eyes. Watching his eyes all the time, first move your hand vertically up and down, then horizontally and finally obliquely. Note the extent of the movement of the eyeball that is possible.

These movements are brought about by six muscles, each of which is attached at one end to the outside of the eyeball and to the orbit at the other. They act like reins guiding the head of a horse; when one contracts and pulls the eyeball in a certain direction, another which is

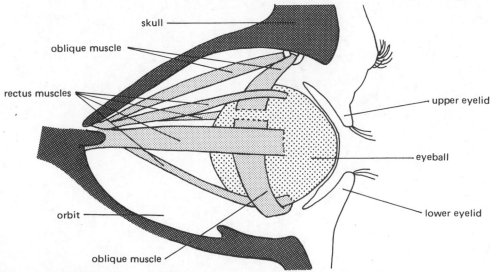

Fig. 8:3 Right eye exposed to show the attachment of the eye muscles.

antagonistic to it, relaxes. Four of these, called **rectus** muscles, act at right angles to each other, the other two are arranged obliquely. Look at Figure 8:3 and work out which muscles are used to bring about the eye movements you have just carried out. You will also find that you can roll the eye, which involves the contraction of all the muscles in a sequence.

Now shut one eye and place your fingers gently on the lid while you roll the other. What movements can you feel in the closed eye? What do you conclude from this?

A sheep's eye

In order to study the structure of the eye we will examine one from a sheep or bullock.

Examine a fresh one. How does it differ from our own? Internally it is very similar.

1. Remove the fat from the back of the eye using a scalpel and a pair of forceps. In life this fat acts as a cushion between the eye and the bony orbit. People who are very ill or who are starving have sunken eyes—can you think why?

2. Look for the stumps of the six muscles which move the eye. Most of their length will probably have been cut away when the eye was removed but you should see where they were attached to the eyeball. Leave the white optic nerve in position; it is a white cord which projects from the back of the eye and takes the nerve impulses to the brain.

3. Try to demonstrate the kind of image produced by the eye by cutting a small window at the back of the eye exactly opposite the pupil (Fig. 8:4). Place a piece of tracing paper over the hole you have made and point the eye towards an electric-light bulb. If the eye is fresh you should see the image on the paper. Is it upright or inverted?

4. Now remove the back third of the eye by making an incision through the sclerotic and cutting right round with scissors. Why is the sclerotic so tough? Separate the two portions,

Fig. 8:4 Eyeball seen from the back to show where cuts should be made.

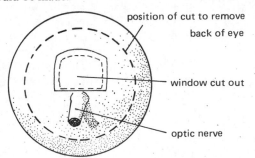

Fig. 8:5 Diagram of a horizontal section through the human eye.

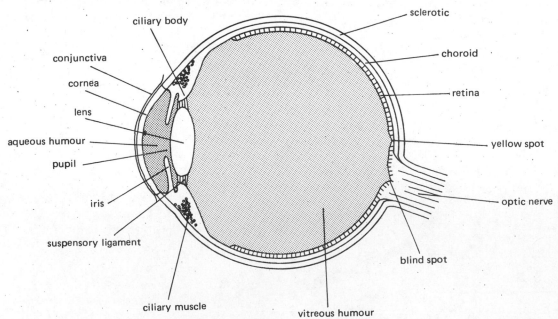

place the back part in a dish of water and examine it. Can you see in the undamaged part very fine lines radiating towards the position of the optic nerve? These are nerve fibres lying on the surface which lead to the optic nerve.

The black colour is due to pigment deposited in the **choroid**, a layer which lies inside the sclerotic. This layer is well supplied with blood vessels and would look pink if it were not for the pigment. What is the purpose of the black pigment? The layer of the eye which is sensitive to light is the **retina**. This is the transparent innermost layer lying on the surface of the choroid from which it receives oxygen and nutrients brought by the blood.

5. Examine the front portion of the eye. Squeeze the eye very gently and note the jelly-like **vitreous humour** which comes out. This will be followed by the lens which should be released very slowly to prevent damage to the tissues by which it is attached. In life the vitreous humour fills the main cavity of the eye and a more watery **aqueous humour** fills the cavity in front of the lens.

6. Examine the lens. If the eye is fresh it will be clear, but it soon becomes opaque after death. (In old age people sometimes suffer from **cataract**. This is a form of blindness due to the lens gradually becoming opaque. It may be cured through the complete removal of the lens. With the help of glasses sight is then possible.)

7. Wash out the front part of the eye and note the ring of **ciliary muscles** to which the lens was attached. Hold the eye up to the light and note the position of the pigmented iris and see how transparent the cornea is. Like the lens, it gradually becomes opaque after death. (If, for some reason, this happens in life it is possible for a surgeon to graft a fresh cornea in its place. People sometimes leave their eyes to a hospital so that when they die these can be used to help somebody who needs this operation.)

8. If available, you should now study an eye which has been in a deep freeze. Cut it lengthwise with a fretsaw, if possible through the optic nerve. This will show you the position of all the parts you have observed and you can compare them with Fig. 8:5. You will now be in a better position to understand how the eye works.

Focusing

When light rays enter the eye they are refracted at the corneal surface as well as by the lens (Fig. 8:6) and are focused on to the retina. A *glass* biconvex lens has a definite focal length and an object is only in focus if it is a fixed distance away; the fatter the lens, the closer this distance becomes. The focal length of the lens of the eye can be adjusted to focus on any object between about 25cm and infinity by altering its shape. This property of adjustment is called **accommodation**.

How is the shape of the lens altered? To understand this it is necessary to know how the lens is attached. If a *fresh* lens is squeezed and then released it will return to its original shape. This is because the lens is surrounded by an elastic membrane which moulds it into a *thick* convex shape, thus when the lens is *released* it has a short focal length (for near objects). In life the lens is attached by a **suspensory ligament** of tough fibres to the ring of ciliary muscles (Fig. 8:6) so it is only when these ligaments are slack that the lens can be in its thick condition. This slackness occurs when the circular fibres of the ciliary muscles *contract*, thus reducing the diameter of the ciliary muscle ring. They contract against the pressure exerted outwards by the fluid contents of the eye. However, when these circular fibres of the ciliary muscles are *relaxed* the ciliary muscle ring becomes larger, due to the outward pressure of the fluid in the eyeball. This pulls on the suspensory ligament which in turn pulls the lens into a thinner shape (focused on more distant objects). Therefore relaxed ciliary muscles result in distant focusing, contracted ciliary muscles result in near focusing. Could this be a reason why prolonged close work is tiring to the eyes?

The retina

The retina contains sensory cells of two kinds, **rods** and **cones**, named because of their characteristic shapes (Fig. 8:7). When these are stimulated by light, electrical impulses pass along the nerve fibres which link them to the brain. These fibres lie on the surface of the retina and converge to form the optic nerve which leads to the brain. So when we look at an object, the brain receives a mass of impulses

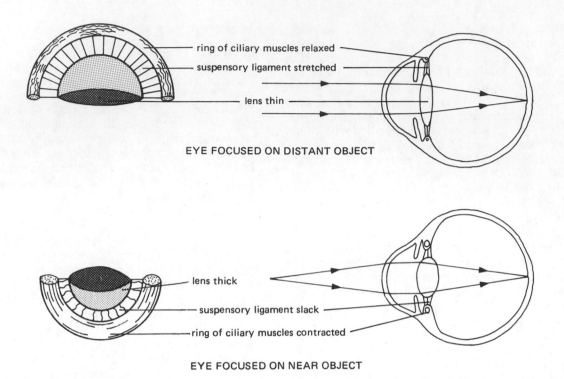

EYE FOCUSED ON DISTANT OBJECT

- ring of ciliary muscles relaxed
- suspensory ligament stretched
- lens thin

EYE FOCUSED ON NEAR OBJECT

- lens thick
- suspensory ligament slack
- ring of ciliary muscles contracted

Fig. 8:6 The mechanism of accommodation: (left) relevant organs as seen from the back of the eye (right) in section.

from the retina and these are interpreted in the form of an image of the object seen.

Rods are more sensitive to light of low intensity and are more suitable for night vision, but the image interpreted by the brain lacks detail; cones, by contrast, need stronger light to activate them but they produce a much more detailed image and are more suited to day vision. Their distribution within the retina is not uniform. Rods are much more concentrated round the periphery while the cones are more numerous towards the centre. The **fovea** or **yellow spot** which is the point of principal focus is composed of cones only. When we look at any object very carefully we place it in just the right place so that it is focused on the fovea. If you look intently at an object on a very dark night you may see nothing, but if you look to one side of it, you may then just see it vaguely because the image does not fall on the cones in the fovea, but on an area nearby which contains rods. What proportion of rods do you think there would be in nocturnal animals such as bats and owls?

Blind spot

Where the optic nerve leaves the eye there are no retinal cells present and so no image is formed. You can demonstrate the presence of this **blind spot** for yourself (Fig. 8:8):

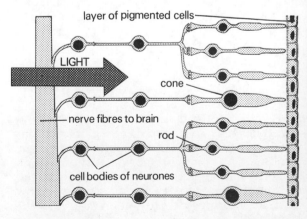

Fig. 8:7 Diagram showing the arrangement of cells of the retina. Note how the light has to pass through the nerve fibres before reaching these cells.

96

●   A   B   C   D   E   F   G

Fig. 8:8

Hold the book at arm's length, close your right eye and look at the black spot with your left; now move your left eye to the right along the series of letters until the spot disappears. What letter were you looking at when it disappeared?

Now hold the book nearer and repeat. Was it the same letter? Repeat for different distances. What do you conclude from this?

Acuity

This term is used for the acuteness of vision, i.e. the amount of detail it is possible to see. If you examine under a lens a photo printed on fine art paper and compare it with one in a newspaper, the latter is seen to be made up of a relatively small number of coarse dots compared with the former which has a larger number of fine ones. In the same sort of way acuity of the eye is dependent upon the number of cones per unit area of retina. The more there are, the more impulses pass to the brain and the more detailed is the image. It also follows that better acuity will result if the eye is large, as the retina will have a relatively greater area for receiving the image. Similarly, if the lens produces a large image it will be spread over a greater number of cones and more detail will be seen. Our eyes have good acuity and we might, for example, observe a car number plate 100 m away and just succeed in counting how many numbers and letters were on it, but at 50 m we might be able to read accurately what the numbers and letters actually were. However, birds have far greater acuity than we have and a hawk or eagle can see details equivalent to our using binoculars with a magnification of eight!

Colour vision

Colour perception in mammals is confined to man and other primates; a bull, for example, does not react to a red flag because it is red, but because it is waved. Most birds, reptiles and fish which have been tested can detect colours, and so can bees and butterflies, but there is variation in the range of colours perceived. Bees can see ultra violet as a colour, but are blind to red. We cannot imagine what ultra violet would look like as ultra violet rays are filtered off by our lenses because they are slightly yellow. In old age this yellow becomes more pronounced with the result that old people have difficulty in seeing some of the violet hues visible to younger people.

It is the cones in our eyes that distinguish colour. As these need light of high intensity to stimulate them, it follows that we cannot see colours in poor light. At dusk everything becomes black or white or shades of grey. It is not known exactly how colour is detected, but put very simply, a likely theory is that there are three kinds of cone which are stimulated by different wavelengths of light corresponding to the red, blue and green parts of the spectrum. Thus the brain builds up a colour picture according to the number of impulses received from the three kinds of cone.

Stereoscopic (binocular) vision

When the eyes are at the side of the head, as in hares and horses, the field of view is very large because both eyes cover different areas and there is little or no overlap. However, in a hare, when its head is facing forwards there is a blind area immediately in front of it, so if the hare's head is pointing towards you it is possible to walk quietly up to it without being seen! In man, however, the eyes are near together and face forward, hence both eyes can see the same object and two separate images are formed. The brain somehow combines these two images so that we do not 'see double'. However, occasionally, if there is an eye defect or if the brain is damaged it is possible to see two images. This may be experienced if you look at an object and gently press one eye upwards. The effect of excessive alcohol and other drugs on the brain may also produce double vision.

The images received by the two eyes are slightly different. This makes the object stand out—we see it in three dimensions—and it also gives us a sense of distance. Test this for yourself with these simple experiments:

1. Hold a ballpoint pen nearly at arm's length and *quickly* try to put the cap on the pen. Now shut one eye and repeat the operation. Which method is more precise?
2. Close your eyes and ask your neighbour to put a small object a few feet away from you at an angle of about 45° from the direction in which your head is facing, keep your head in the same position and open your left eye and estimate how far away the object is; repeat for the right eye only; finally open both and estimate the distance once more.

Measure the distance and see which estimate is the most accurate.

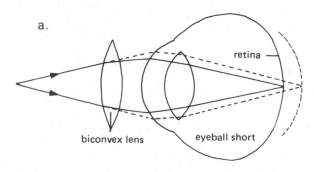

a.

retina

biconvex lens   eyeball short

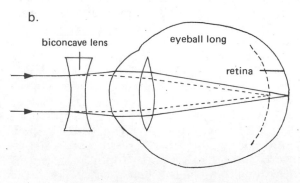

b.

biconcave lens   eyeball long

retina

Fig. 8:9 a) Lens corrections for long sight. Dotted lines indicate that focus of near object is behind retina; this is remedied by a biconvex lens. b) Lens correction for short sight. Dotted lines indicate that focus of distant object falls in front of retina; this is remedied by a concave lens.

How important do you think binocular vision would be to: 1. A mammal which eats grass, 2. A carnivore which feeds on small rodents, 3. A mammal which leaps about in trees? Test your answer by thinking of different animals in these categories and see how their eyes are placed in their heads.

Correction of faulty vision

The most common kinds of faulty vision are long and short sight. When a **long-sighted** person (Fig. 8:9a) views a near object the image falls behind the retina. This is corrected by wearing glasses with convex lenses which help to converge the light rays before they enter the eye. In middle age many people who have had excellent eyesight begin to find reading difficult; as they get older they have to hold a book further and further away in order to focus. This is because the lens gradually hardens with age, it loses some of its elasticity and so its power of accommodation is reduced. Glasses with convex lenses correct this.

Short-sighted people can see objects in focus near the eye, but with long distance the image is focused in front of the retina. This is usually the result of the eyeball being too long (Fig. 8:9b). This is corrected with concave lenses which diverge the light rays before they reach the eye. Some forms of short sight improve with age.

Astigmatism is a condition where the cornea is uneven; it refracts the light rays in an abnormal way and so produces a distorted image. A regular astigmatism occurs when the cornea has a different curvature in the horizontal and vertical planes; this type of defect can be corrected with the help of glasses having a lens of appropriate shape to compensate for the distortion.

THE EAR

The ear provides the brain with information about the sounds around us, their pitch and loudness and the direction from which they come, but the ear is also an organ of balance and helps in maintaining posture.

Its structure is shown simply in Fig. 8:10. It is divided into three regions—an outer, a middle and an inner ear— all three parts being concerned with hearing, but only the inner ear with balance and posture.

Outer ear

This consists of the part you can see, the **pinna**, and a tube which passes into the skull and ends at a membrane, the **tympanum** or **ear drum**. The pinna helps to concentrate sound waves and direct them towards the tympanum, but in man, because the pinna is immovable, it is not nearly so efficient as in most mammals. A deaf person may cup his hand behind his ear to improve its collecting capacity and at one time ear trumpets were used for the same purpose.

Just as with binocular vision, when two eyes are focused from different parts of the head on a single object making it possible to judge its distance, many animals with their two ears can judge the position from which a sound comes by moving them independently until each receives the maximum sound. In this way cats and foxes can leap on a mouse in the dark, judging its position by sound. In the serval cat this has become so highly developed that its huge ears act like radar screens, magnifying the slightest sounds and enabling it to leap on a hare feeding unseen in long grass from a distance of at least 3 metres.

Our ears cannot move, but we can locate the direction of a sound nevertheless; this is because the sound is heard more loudly by the ear nearest to it and also fractionally earlier. A sound is always difficult to locate if it comes from a source equidistant from both ears. That is why we turn our heads to one side to make sure of the direction from which the sound is coming.

Test this for yourself:

> One person is blindfolded and asked to indicate the position of a ticking clock by pointing in the direction from which the sound is thought to come. He must keep his head still all the time. The clock should be held in ten positions where it is equidistant from both ears and in ten positions where it is at different distances. How often is the estimate of direction correct for each category?

The middle ear

This is an air-filled cavity surrounded by bone. Its main function is to transmit the sound vibrations which cause the tympanum to vibrate to the inner ear by means of three small bones called the **hammer**, **anvil** and **stirrup**, so named because of their characteristic shapes. The hammer touches the tympanum and the stirrup is in contact with a membrane covering the **oval window** of the inner ear. Because this membrane is much smaller than the tympanum the force of the vibration it receives is much greater; it is further increased by the lever action of the ear bones with the result that the inner ear receives sounds amplified about 22 times.

The only opening of the middle ear to the outside world is via the **Eustachian tube** which opens at the side of the throat. Usually this opening is kept closed, but when we swallow or yawn it opens. It is through this opening that air can pass, thus keeping the pressure equal on both sides of the tympanum. We notice this when gaining height rapidly in an aircraft or even when we are going up a long steep hill in a car: we feel our ears pop as equal pressure is restored. Why would the chewing of sweets help to prevent this from happening?

If a person has a sore throat due to some bacterial infection the Eustachian tube is a possible route through which the infection may spread to the middle ear and cause ear ache.

The inner ear

This is the part of the ear where the sensory cells are situated and from which impulses pass to the brain via the nerves. It consists of the **labyrinth**, a delicate hollow structure filled with fluid (**endolymph**) and surrounded by more fluid (**perilymph**); it is deeply embedded in bone. Its parts have different functions; the **cochlea** is the organ of hearing and the **sacculus**, **utriculus** and **semi-circular canals** are for maintaining balance and posture.

The cochlea is a much-coiled tube with a blind end. In it there is a long ribbon-like membrane which passes along its length. This membrane is composed of transverse fibres of varying lengths. Vibrations received at the oval window are transmitted through the fluids of the cochlea causing the transverse fibres of the membrane to vibrate at certain places according to the frequency. High notes cause the short fibres of the front part of the membrane to vibrate, low notes stimulate the

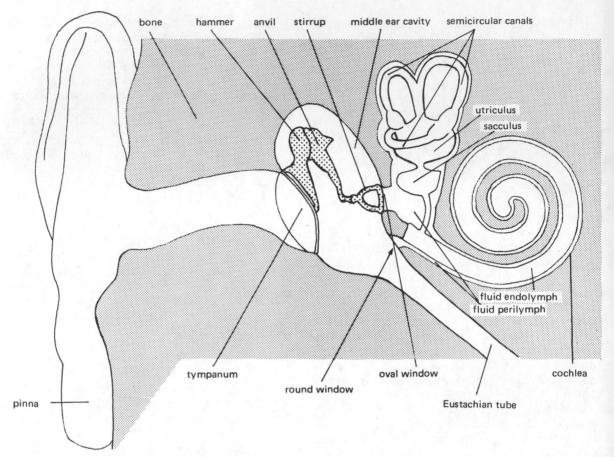

Labels on figure: bone, hammer, anvil, stirrup, middle ear cavity, semicircular canals, utriculus, sacculus, fluid endolymph, fluid perilymph, cochlea, oval window, round window, Eustachian tube, tympanum, pinna

Fig. 8:10 The human ear

longer fibres towards the far end. These fibres are in contact with cells bearing sensory hairs, which in turn have nerve connections to the brain. In this way the brain receives information concerning the sound received, according to which sensory cells are stimulated.

As the fluid in the cochlea is virtually incompressible there has to be another membrane which can vibrate and compensate for changes in pressure; such a membrane is present covering the **round window** which thus vibrates in sympathy with the oval window membrane. The sequence of events can be summarised as follows:

vibrations in air ⟶ tympanum ⟶ ear bones
— fluids in cochlea ⟵ oval window ⟵
membrane in cochlea ⟶ hair cells $\xrightarrow{\text{nerve impulses}}$ brain

Range of sounds detected

When young we can hear sounds ranging from about 60 Hz to 20,000 Hz. One hertz is a single to-and-fro vibration per second. But as we get older we find it more difficult to hear very high notes such as the squeak of a mouse or bat; this is because our tympanum becomes thicker and our ear bones do not transmit high frequency vibrations so well. We are most sensitive to sounds of a frequency of about 3000 Hz. It is interesting that this corresponds to the piercing sound of a child or woman screaming for help. Other mammals are most sensitive to different ranges. A cat, for example, is sensitive to the very high ranges such as squeaks made by mice, but mice can hear still higher warning noises from other mice which a cat cannot hear! Bats, whales and probably many other species communicate with higher frequencies still, far beyond our powers of hearing; they also use these sounds for echo-location (Book 1 p. 114).

100

Find out your own range of audible sounds by using an audio-signal generator. This gives out sounds of known frequencies over a wide range through a loudspeaker. As each sound of a particular frequency is emitted record on paper whether you can hear it or not. At the end you will be told the actual frequencies which were transmitted. Consider a) whether or not it is important, in obtaining accurate results, for the distance between each person and the loudspeaker to be roughly the same, b) whether it is better for the sounds to be emitted in a regular sequence from lower frequency to higher or in a more random manner.

How much variation in hearing ability is there in your class?

Does sound reach us only through our outer ear?

Put your first fingers into both your ears and start to hum. Can you hear the sound? Keep putting your fingers in and out. Do you hear the sound more strongly with your fingers in or out of your ears?

When you hum, your lips are closed and most sound passes to the inner ear through the skull itself. Under normal circumstances we hear the sound of our own voice via our inner ear and through our skull, but another person only hears our voice through the outer ear, that is why we hardly recognise our own voice when we play it back on a tape recorder, but other people do.

Even external sounds are transmitted to some extent through the skull. A hearing aid placed just behind the ear works on this principle; it amplifies the vibrations it receives from the air, and because it is in contact with the skin which covers the bone the hearing aid enables the vibrations to be transmitted to the inner ear.

Balance and posture

In the labyrinth there are structures which respond to gravity in such a way that when our head is tilted we are made aware of its position even if we are blindfolded and our head is quite still. In the wall of the utriculus there are special groups of cells with fine projecting

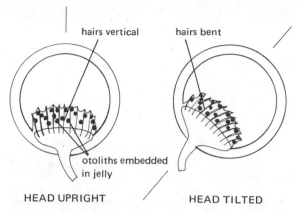

Fig. 8:11 Diagram illustrating otolith action.

hairs; particles of calcium carbonate called **otoliths**, embedded in jelly, make contact with these hairs (Fig. 8:11). When the head is upright the otoliths press downwards on the hairs due to gravity. If we stand on our head, the pressure on them is much reduced and if we tilt our head the otoliths bend the hairs to the side. As a result of these various stimuli the brain obtains information about the head's position and responds by causing appropriate muscles to contract which bring the head back to its normal position. Thus posture can be adjusted quite unconsciously.

In addition to these organs which are concerned with posture there are others which respond to directional movements; these are the semi-circular canals. There are three of these arranged in three planes at right angles to each other, two vertical and one horizontal. One end of each canal is swollen and contains a group of sensory cells whose projecting hairs are embedded in a cone of jelly. When we move our head from a resting position in a particular direction the fluid inside the canal which lies in the same plane as the movement exerts pressure on the cone of jelly, thus bending the hairs. As a result impulses are sent along nerves to the brain. In this way the brain receives information about **accelerating** movements in a particular plane or **rotational** movements. If speed is constant in one direction pressure on the hair cells will be regular and no further impulses will be received by the brain. You should now be able to work out what happens when you twirl your body round rapidly and stop suddenly. What kind of sensation do you feel?

SENSE ORGANS WHICH RECEIVE CHEMICAL STIMULI

The olfactory organs

These are the organs of smell and consist of sensory cells which line the roof of the nasal cavities. When we take in air through our nostrils many molecules of volatile substances pass in with it; these dissolve in the mucilage covering the sensory cells, stimulating them and causing impulses to pass to the brain. We also smell substances which we take into our mouth, as molecules from them can pass into our nasal chambers via the back of the throat (Fig. 8:12). We usually say that we can taste them, but really we are smelling them. When we have a bad cold this back entrance to the nasal cavities may become blocked with mucus and we can no longer detect the flavour of our food.

We have a very poor sense of smell compared with most mammals, but even so most of us can easily distinguish over 1000 different odours while some people can distinguish up to 4000.

The organs of taste

These are called taste buds, which are groups of cells situated on the tongue (Fig. 8:13). As we have seen, these do not detect flavours (which are smelt), but are sensitive only to substances which are sweet, sour, salt or bitter. Their chief function appears to be to give warning regarding the suitability or unsuitability of food before it is swallowed.

The distribution of taste buds on the tongue is not even. You could map out the position of those concerned with the four sensations in this way:

Use solutions of sugar (sweet), very dilute hydrochloric acid (sour), table salt (salt) and quinine (bitter). Work in pairs and use one solution at a time. One partner should put his tongue out while the other transfers a drop of solution on the end of a glass rod to one of the four regions of the tongue to be tested—back, tip, centre and sides. Repeat for the three other regions, rinsing out the mouth between each application. The tongue should be kept

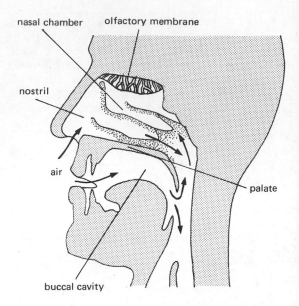

Fig. 8:12 The organs concerned with smell.

still while each drop is being tested. Record which regions can detect each solution.

SENSE ORGANS WHICH DETECT TOUCH, PRESSURE, PAIN AND TEMPERATURE

Receptors in skin and muscles

These are mainly situated in the skin and are of several kinds, each concerned with a particular stimulus. However, it does appear that in some cases a number of different stimuli can be detected by the same sense organ; this is particularly true of sensory nerve endings which register pain, but are also sensitive to touch and temperature (Fig. 8:14). Can you think why it is advantageous for the brain to monitor pain? Although widely distributed over the skin some areas have concentrations of sensory cells of a particular kind, making such places particularly sensitive; thus the lips and the tips of the fingers are very sensitive to touch and the upper arm to temperature. This is why a mother will often judge the temperature of a baby's bath by putting her elbow rather than her hand in the water. An object feels warm because our sensory cells detect a flow of energy from the object to us; it feels cold when the flow goes in the other direction. Test this for yourself:

102

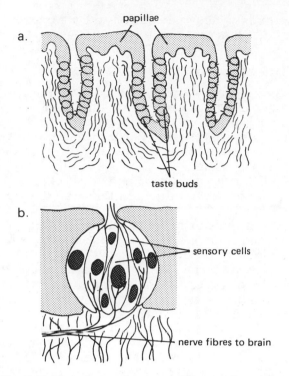

Fig. 8:13 The organs of taste: a) diagram of a section through the upper surface of the tongue b) a taste bud much enlarged.

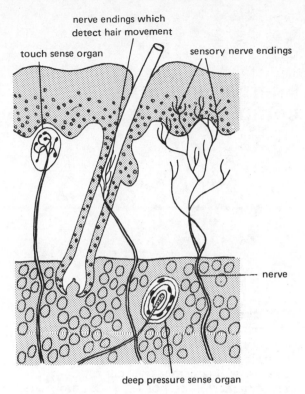

Fig. 8:14 Diagram of a section through human skin showing sense organs.

Put one hand in ice-cold water and another in warm water. Now touch a stone which is at room temperature with each hand in turn. Which hand's reaction are you going to believe?

In muscles and tendons there are nerve endings which act like strain gauges, sending stimuli to the brain according to the degree of stretching. This helps co-ordination of movement and is another means of maintaining posture (p. 90).

9

Internal lines of communication

We have seen in the previous chapter how information is constantly being received by the sense organs about the changes which are occurring both outside and inside our bodies. Now we will consider how appropriate action may be taken in response to the information received.

There are two linking or co-ordinating systems between the **receptors** which detect the stimuli and the **effectors** which react in consequence; these are the nervous and endocrine systems. The nervous system co-ordinates the activities of the body by means of a complex system of nerves, while the endocrine system does so by secreting hormones into the blood.

THE NERVOUS SYSTEM

This consists of:
1. The **central nervous system** (CNS), composed of the brain and spinal cord. This part of the system co-ordinates the impulses received from the receptors and transmits other impulses to the effectors which then act in response.
2. The **peripheral nervous system**, composed of paired **cranial** nerves coming from the brain and paired **spinal** nerves from the spinal cord. These nerves are the living lines of communication between the receptors, the central nervous system and the effectors.
3. The **autonomic nervous system** which is concerned with the body's automatic (involuntary) activities, such as the contractions of the alimentary canal and the beating of the heart.

The neurone

The nervous system is made up of units called nerve cells or **neurones**. There are three kinds:
1. **Afferent** or **sensory** neurones which transmit impulses from a receptor to the central nervous system.
2. **Efferent** or **motor** neurones which transmit impulses from the central nervous to the effectors.
3. **Association** neurones which link the afferent and efferent neurones; these lie within the brain or spinal cord.

Each neurone (Fig. 9:1) consists of a **cell body** composed of cytoplasm and a nucleus and a number of cytoplasmic extensions. Of the latter, one is a long thin process called an **axon**, while the others are shorter and end in many fine **dendrites**. Neurones include some of the longest cells in the body because their cell bodies are situated mainly in the brain and spinal cord, but their axons may reach the furthest extremities of the body.

The cytoplasm of one neurone is not continuous with that of another, but the dendrites of one become very closely associated with the terminal branches of another neurone or with its cell body and the impulse is able to pass across the gap. These gaps between interconnecting neurones are called **synapses**.

The nerve impulse

A neurone is able to transmit an electrical impulse very rapidly. The impulse is not the same as an electric current passing down a wire but takes the form of a wave of electrical disturbance along the neurone. The impulse normally travels along a particular neurone in one direction only and it does so at a speed of up to 100m/s in man.

Nerves

These are aggregations of axons (collectively called nerve fibres) bound together like wires in a cable (Fig. 9:2). The fibres of cranial and spinal nerves are insulated from each other by a fatty sheath and are called **medullated** fibres.

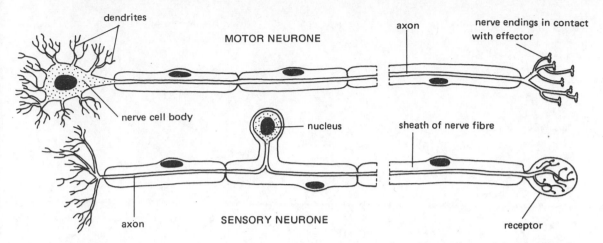

Fig. 9:1 Diagram showing the structure of a motor and a sensory neurone. They are usually much longer than shown.

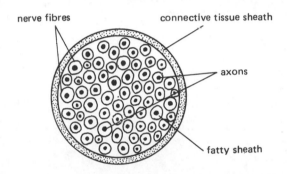

Fig. 9:2 Diagram of a transverse section through a small nerve.

Simple reflex actions

These are the simplest examples of nervous co-ordination. We have already come across some of the simple reflexes which are associated with keeping the head upright. You will remember that stimuli received by the eyes, ears and 'strain gauges' in the neck muscles (pp. 101 and 103) all bring about compensating head-righting actions. A doctor can test your reflexes in various ways. In one method he asks you to sit on a stool with your right leg crossed over your left so that it hangs freely; he will then tap the tendon just below the right knee cap. If the reflex is working correctly, the leg jerks forward. If there is no response the doctor will know that there must be something wrong with the nerves of the leg or the spinal cord itself.

Test your neighbour's reflexes in a similar manner. Is it possible to prevent this response?

There are many other simple reflexes:
When you touch something hot the finger is quickly withdrawn from the stimulus. When a light is shone in your eye, the pupil contracts. When pepper gets up your nose, you sheeze. When a crumb goes down 'the wrong way', i.e. enters the glottis, you choke.

Can you think of other examples?

Simple reflexes have these important characteristics:

1. They are inherited, so they do not have to be learnt and are not forgotten.
2. They are not under the control of the will and so are quite automatic.
3. For a given stimulus, the response is always the same.

The reason why the last statement is true is because the impulse travels along the same nervous pathway. This pathway is called a **reflex arc**. In its simplest form, as in the knee jerk reflex, each reflex arc consists of two neurones, an afferent and an efferent, which associate at a synapse in the central nervous system. However, most reflex arcs (Fig. 9:3) have, in addition to the afferent and efferent neurones, one or more association neurones which enable a stimulus such as a prick, received at a single point, to be transmitted via

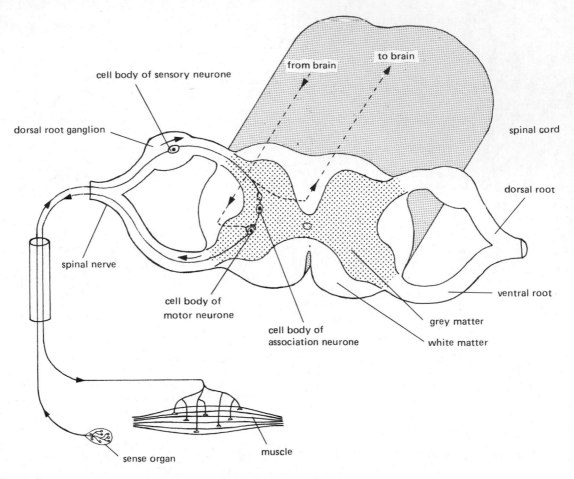

Fig. 9:3 The structures concerned in a reflex arc involving the spinal cord.

many efferent neurones, causing whole systems of muscles to move the complete arm. At the same time, other neurones send impulses along the spinal cord to the brain and so make us conscious of the prick. Impulses can also be sent from the brain down the spinal cord, and these co-ordinate the action of motor neurones in other spinal nerves. This allows the brain to control any further action which is necessary. For example, sometimes, when we inadvertently touch something hot, we draw the hand away by a reflex action, but if immediately afterwards our brain tells us the object was not hot enough to cause pain we may consciously put our finger back on to it again.

Reflex arcs involving sense organs in the head pass through the brain and are called **cranial reflexes**, e.g. we automatically blink when an object approaches the eye. Those which involve sense organs from the neck downwards pass through the spinal cord and are called **spinal reflexes**. In some animals it is possible for the latter to continue to work for a time after the brain has been destroyed.

From the brain arise 12 pairs of **cranial** nerves and from the spinal cord 31 pairs of **spinal** nerves. The latter emerge through holes between adjacent vertebrae in a very regular manner down the spine (Fig. 9:4).

You will see from Fig. 9:3 that the spinal nerves have a dorsal and a ventral root, and that afferent neurones pass into the dorsal root, their cell bodies being confined to a swelling, the **dorsal root ganglion**. The efferent neurones have their cell bodies in the spinal cord itself and their axons pass out along the ventral root.

106

The central nervous system

The spinal cord looks fairly solid in section (Fig. 9:3) but it is in fact a hollow tube, as there is a canal in the centre which is continuous throughout its length. This canal contains cerebro-spinal fluid and is continuous with certain cavities in the brain called **ventricles**.

The brain can be looked upon as a highly specialised part of the spinal cord. In the embryo it arises as three swellings at the anterior end of the spinal cord, each part differentiating later into the main structures of the adult brain.

Both brain and spinal cord are made up of nervous tissue of two kinds, **grey matter** and **white matter**. The grey matter consists largely of nerve cell bodies, and the white matter of nerve fibres surrounded by their medullary sheaths which cause the white appearance. In the brain the grey matter tends to be situated near the outside; in the spinal cord it is confined more to the centre.

The brain

As the brain develops from the anterior part of the spinal cord, it is not surprising that it has rather similar functions; however, these are carried out on a much more complex scale. Like the spinal cord, the brain receives impulses from sense organs, especially the nose, eyes and ears. It also sends impulses to the muscles and glands of the head, and to those in other parts of the body via the white matter of the spinal cord. The brain differs greatly from the spinal cord, however, in the astronomical number of association neurones that it contains. These allow an almost infinite number of cross-connections to occur, and consequently great variation in behaviour is possible. How are all these impulses sorted out? How are the pathways for the stimuli determined? What makes the brain act as a completely co-ordinated structure and not as a lot of isolated centres? What actually happens when we 'make up our mind' about something? What process is involved when we learn? How is information stored so that we can act according to our experience? The answers to these intriguing questions are only partially known, but brain research is providing useful clues towards the solution of some of them. A brief study of the structure of the brain should help you to understand some of the principles involved. Let us consider in particular the cerebrum, the cerebellum and the medulla (Fig. 9:7).

The cerebrum

In the course of evolution of the vertebrate brain the cerebrum has enlarged more and more and in man forms by far the largest part of the brain. It consists of two lobes or **cerebral hemispheres** partially separated by a deep cleft in the midline. The two hemispheres are connected at a deeper level by the **corpus callosum**, a broad sheet of white fibres which helps to co-ordinate the left and right sides.

The surface region, called the **cerebral**

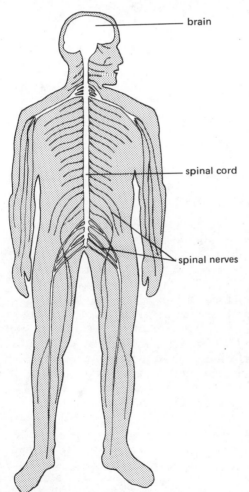

brain

spinal cord

spinal nerves

Fig. 9:4 Diagram of the central nervous system and the spinal nerves of man.

107

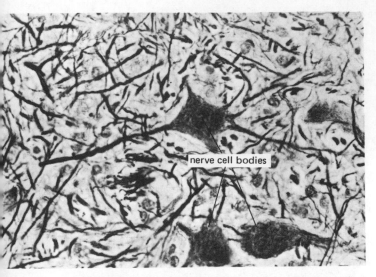

Fig. 9:5 High power photomicrograph of brain tissue showing neurones.

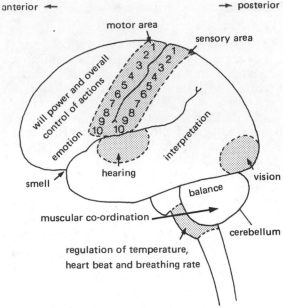

Key to corresponding motor and sensory areas:

1. foot 2. leg 3. trunk 4. arm 5. hand

6. fingers 7. eye 8. face 9. lips 10. tongue

Fig. 9:6 Side view of human brain showing the main regions.

cortex, is thrown into intricate folds and grooves which give it a far greater surface area. This is important because the cortex consists of grey matter and the more there is of it, the more nerve cell bodies it can contain. In fact, about 90% of all the nerve cell bodies in the brain occur here.

The cerebral cortex is by far the most important **association centre** of the brain. All the time information from our main sense organs streams into it and is sorted out in the light of our past experiences, as a result of which motor impulses are discharged from it along pathways of white fibres and cause the appropriate action to be taken. When learning takes place it is possible that certain neurones may become associated in some way to form a particular pathway which may be followed once more if the same stimulus is received. But the exact processes of learning and storing memories are not yet understood.

When certain parts of the brain are given small electric shocks, particular sensations are felt in the body or different muscular movements occur. In this way it has been discovered that specific areas of the cortex deal with skin sensations from various parts of the body and other areas are concerned with the control of muscular movements in different regions. Figure 9:6 shows how these sensory and motor areas form bands across the cerebrum and how other regions have different functions.

The cerebellum

This region is concerned with balance and posture and co-ordinates muscular movement so that all the appropriate reflexes take place at the same time. When a child is learning how to walk, its movements are at first unco-ordinated; similarly, when we learn to ride a bicycle, drive a car, or play the piano there are many intricate movements to be co-ordinated. Muscular control of these movements is performed by the cerebellum, although, as you would expect from the fact that these activities have to be learnt, there are connections between it and the cerebrum.

The medulla

This forms the brain stalk and links the rest of the brain, by means of large numbers of nerve fibres, with the spinal cord. In it are the centres which control automatically such vital functions as rate of breathing, regulation of temperature and rate of heart beat.

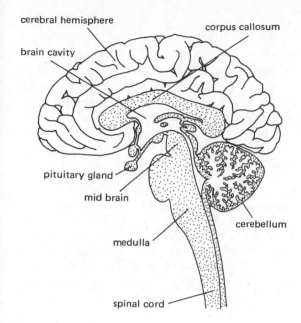

Fig. 9:7 Vertical section through a human brain.

Labels on Fig. 9:7:
- cerebral hemisphere
- corpus callosum
- brain cavity
- pituitary gland
- mid brain
- cerebellum
- medulla
- spinal cord

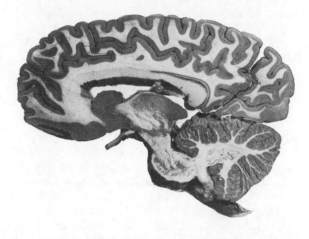

Fig. 9:8 Human brain: photograph of a similar section to the one shown in Fig. 9:7.

Circulation of blood to the brain

There is an excellent blood supply to the brain as the nerve cells require a constant supply of glucose and oxygen. If the blood supply becomes inadequate for some reason, **fainting** occurs. This sometimes happens when soldiers are on parade because they are standing very rigidly for a long time with very tense muscles. A person who has fainted should be placed so that the head is low in relation to the rest of the body in order that gravity may aid the pumping action of the heart and help the circulation to the brain. As a result of age or injury, small arteries in the brain may become damaged and so cause local haemorrhage, or a blood clot may form within an artery. In both cases, brain tissue will be damaged. This is the cause of a **stroke** which, if severe, may cause paralysis or death.

The autonomic nervous system

This system is concerned with many of the actions which go on within the body quite unconsciously. It has two divisions, **sympathetic** and **parasympathetic**.

Sympathetic fibres leave the brain as a pair of nerve chains which pass down the body on either side of the vertebral column. On these chains are ganglia which contain the cell bodies and axons of sympathetic neurones; some of these neurones transmit stimuli to various internal organs while others make links with the spinal cord. A further group of ganglia occurs in the abdominal region and comprises the **solar plexus** from which nerves pass to the gut and other abdominal organs. A blow in this region can produce a violent and painful effect.

Parasympathetic fibres come mainly from the base of the brain and pass to the organs via the tenth pair of cranial nerves.

Sympathetic and parasympathetic fibres supply the same organs, but their effects are usually antagonistic. This provides a means of regulating an action according to changing circumstances. For example, if the sympathetic nerve fibres to the heart are stimulated the heart beats faster; if the parasympathetic fibres are stimulated the heart rate is depressed. In a similar way, stimulation of sympathetic nerves causes the blood vessels of the skin and gut to constrict and the iris of the eye to dilate, while stimulation of parasympathetic nerves produces the opposite effects. These effects all concern the action of **involuntary** muscle—the kind of muscle found in the walls of the gut, uterus and blood vessels—the contractions of which are slow and rhythmical. In addition, the autonomic nervous system controls the secretion of glands, such as the salivary and sweat glands, and such endocrine organs as the adrenal glands. In general terms, it may be said that when sympathetic nerves are stimulated, the overall effect is to mobilise the body for instant emergency action, while the stimulation of the parasympathetic nerves have a calming effect on the body.

THE ENDOCRINE SYSTEM

This is the second linking system. We have already seen that certain hormones are concerned with the regulation of metabolic processes (p. 75) and that hormones from the pituitary and thyroid glands affect growth and reproduction (p. 25). Some of the reproductive hormones also have a co-ordinating effect such as those concerned with the ovarian cycle. The first hormone to be discovered was named **secretin**; this too has a co-ordinating function. When food enters the duodenum from the stomach certain of the lining cells, as a result of coming in contact with food, secrete this hormone into the blood stream. During its circulation round the body the only organ which the hormone affects is the pancreas which is stimulated to secrete pancreatic juice, thus ensuring that the enzymes reach the food at the right time.

Other important co-ordinating hormones are secreted by the adrenal glands.

Adrenal glands

These are paired structures situated near the kidneys. They consist of two parts, each of which secretes quite different hormones. The outer part—the **cortex**—secretes several which are concerned with salt and water balance, maintenance of blood pressure and resistance to stress; the inner part—the **medulla**—secretes **adrenaline**, an important co-ordinating hormone.

Adrenaline is secreted when we become suddenly frightened. We know the feeling only too well; our heart starts thumping, we get a sinking feeling in our stomach, we break out into a sweat, our skin goes pale, our pupils dilate and it is even possible, sometimes, for our hair to stand on end. All these reactions are caused by adrenaline, and their combined effect, along with other internal reactions which we cannot feel, prepares the body for such actions as 'fight or flight'. In such an emergency the body suddenly needs a large supply of energy for the muscles and many of the fright symptoms we experience are connected with this action:

1. The heart beats faster and more strongly, so blood reaches the muscles more quickly.
2. The walls of many small arteries constrict; those in the skin cause paleness, those in the gut give us that sinking feeling. These actions combine to direct the blood to the parts which need it most, especially the muscles, the brain and the lungs.
3. The breathing rate increases, so more oxygen passes into the blood and reaches the muscles, and more energy can be released.
4. More glycogen in the liver is broken down into sugar, the level of blood sugar rises, and so more is available for energy release.

In animals such as cats, the raising of the hair and the dilation of the pupils help to give the impression of greater size and fierceness; this might make the aggressor hesitate before attacking.

Adrenaline is also secreted when we are nervous or worried. It causes the rather unpleasant feeling we have when we are waiting for an interview, or for a match to begin, or when we are called to do something in front of a lot of people. It is also secreted when we have a bad dream and wake up with our heart thumping away and our skin sweating. Fortunately, the effects of this hormone do not last long as adrenaline is quickly destroyed after the emergency. However, adrenaline is not only secreted when there is an emergency; it is passed into the blood stream in small amounts all the time. But too much regular secretion of adrenaline over a long period, as may occur when people are living under constant stress, may cause the heart to become overworked.

You will have noticed the similarity between the effects of adrenaline and those of the sympathetic nervous system. This is because the two actions are closely linked and one enhances the other. The action of adrenaline is somewhat intermediate in character between hormonal and nervous co-ordination. Most hormones are slow in their action and spread their influence over much longer periods of time, they are also more concerned with the regulation of metabolic processes. Nervous co-ordination, on the other hand, is largely concerned with momentary stimuli, action is more rapid and the response is more specific. Consequently, hormonal and nervous co-ordination both ensure that all the organs and organ systems of the body work together in an appropriate manner and at the right moment. This enables the body to behave as a whole organism rather than as a collection of independent parts.

THE EFFECT OF DRUGS ON THE NERVOUS SYSTEM

The term drug is used for any chemical which alters the functioning of the body. In medicine drugs play a vital part in saving life and combating ill health. Some drugs specifically affect the nervous system and these range from the comparatively harmless caffeine in tea and coffee to dangerous killers such as heroin, cocaine and morphine. It is the more dangerous ones that concern us here. Their effects on the nervous system vary greatly, but they may be divided into:
1. Sedatives or depressants, e.g. alcohol, barbiturates and other sleeping pills.
2. Stimulants or pep pills, e.g. amphetamines.
3. Mood changers and those which cause hallucinations (illusions), e.g. cannabis and L.S.D.
4. Pain killers, e.g. heroin and morphine.

While some of these drugs, when given under medical supervision, are very beneficial in treating certain conditions, their use can be greatly abused and much harm may result. Normal, healthy people do not need drugs.

Some drugs temporarily produce pleasurable sensations and many people experiment with them for that reason. This is dangerous because it may lead to **dependence** on that drug or to experimentation with much more lethal kinds. A person is incapable of knowing whether he is likely to become dependent until this state has been reached. It is then often very difficult for a cure to be effected. You may know people who have become dependent upon drugs such as nicotine and alcohol and are unable to give them up.

Dependence on a drug may be of two kinds:

1. Psychological dependence
This may have a social origin, for example when a person depends on a drug in order that he may fit into a particular kind of life. He may be over-anxious, shy or feel incapable of behaving like others, and so becomes dependent on drugs to help him. When the effect wears off he suffers a marked reaction and feels the need for more. Dependence may also occur when a person comes to rely on a drug to provide enjoyment. This is characteristic of people who have not learnt to enjoy normal, healthy pleasure or make satisfying friendships. They crave for the extreme sensations which some drugs produce and after stimulation find life utterly boring. This may lead to vandalism and violence. Others become dependent because they cannot face up to personal problems and use drugs as a way out.

2. Physical dependence
This occurs when a person becomes dependent on a drug because, without it, painful physical symptoms develop and these become worse and worse. This is characteristic of the heroin addict.

The effects of some of the common drugs can be summarised as follows:

Alcohol
Its main action is to affect the higher centres of the brain. This may have the effect of removing shyness and self-criticism, and by removing inhibitions lead on to a feeling of bravado and a consequent loss of self-control. The sense of judgement is greatly impaired and reactions become slower. Alcohol also affects the regions of the brain concerned with speech, which becomes slow and blurred in consequence. Co-ordination of movement is impaired and this accounts for the staggering gait of the drunk. Alcohol is absorbed quickly through the stomach wall and when large quantities are consumed over a short period unconsciousness and sometimes death may result.

There are several fallacies commonly held about the action of alcohol. It is not a stimulant, but slows down or inhibits brain action. It does not give you strength, although it may give you the illusion of it. It does not help you to concentrate, but makes you more muddled. It does not make you warmer although it makes you *feel* warmer.

There is a considerable variation in the effect that a given amount of alcohol has on different people. Those not used to it may be affected to

Fig. 9:9 Cirrhosis of the liver—a condition resulting from excessive drinking of alcohol. The normal liver cells are destroyed and replaced by fibrous tissue.

a greater degree; this is also true if it is taken when the stomach is empty.

The habit of drinking alcohol can lead to addiction. People dependent on it are called **alcoholics**. It is estimated that there are about 400,000 alcoholics in Britain alone and over 5 million in the U.S.A.

The social implications of too much alcohol include:

1. A breakdown of human relationships, often resulting in broken homes and mental illness.
2. An increase in the number of road accidents.

Even a small amount can affect judgement and speed of reaction and, although individuals differ in their reactions to quantities of alcohol, all are seriously affected by a level of 0·08% in the blood. This is the level above which people can be convicted for driving under the influence of alcohol.

3. An increase in the number of unwanted pregnancies due to the breakdown of inhibitions and self-control, and the greater liability of seduction.
4. An increase in violence, anti-social behaviour and crime.

Nicotine

We have referred to the effect of nicotine in tobacco smoke on the circulatory system and its consequent effect on heart disease (Book 1 p. 143), but as a drug it also affects the nervous system, acting both as a stimulant and a depressant according to circumstances. Dependence on nicotine is easily acquired, and most smokers experience difficulty in giving up the habit.

Barbiturates

These are sedatives and are used as sleeping pills. In larger doses their effects are very similar to alcohol, slowing down the brain's action and affecting judgement and co-ordination.

The amphetamines

These are the usual ingredients of pep pills. They cause nervous excitement and sleeplessness, enabling a person to avoid having to sleep, even if very tired. However, after the effect has worn off, it is followed by extreme fatigue, depression and irritability. Habitual use leads to lack of concentration, poor health and inability to cope with a job.

Cannabis (or pot or marijuana)

This affects the mood of a person. It acts at first as a stimulant, producing a feeling of elation; self-confidence is increased and this often leads to irresponsible action. Different people react to the drug in various ways; usually there is a change in perception, so that tastes appear different, time appears to slow down, colours may appear more vivid and still objects may appear to be moving rhythmically. After a time its effect changes and it acts as a depressant, producing drowsiness and sleep.

Research workers differ in their conclusions as to whether cannabis damages the brain cells permanently or not. One of its dangers is that it may lead to the taking of much more serious drugs such as L.S.D. or heroin.

L.S.D. (or **acid**)

This is a very dangerous stimulant which may damage the brain permanently if taken regularly. It causes hallucinations (illusions) which have resulted in murders and suicides. It may cause grave emotional reactions and lead to serious mental illness.

Heroin

This drug is taken by mouth or by injection. It is extremely dangerous as it is highly addictive and as the body becomes used to it, more and more is needed to satisfy the craving. If the dose is delayed 'withdrawal symptoms' occur; these include pain in the limbs and abdomen and violent twitching accompanied by hallu-

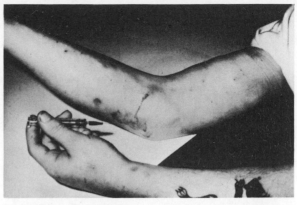

Fig. 9:10 A heroin addict injecting himself.

cinations and nightmares. These conditions are so unpleasant and alarming that the addict becomes more and more afraid of not being able to get the next 'fix' to relieve the symptoms for another brief period. This process frequently leads to criminal behaviour in an attempt to obtain more heroin.

Addiction is a tragic condition and leads to complete breakdown of normal life; typically, death occurs in the mid-thirties. In 1972 there were between 2200 and 2400 heroin addicts in Great Britain.

10

Behaviour

We can define behaviour as the observable reactions of an organism to changes in its environment. These changes may come from outside the body, such as changes in temperature, humidity or light intensity, and also from inside, such as feelings of hunger or pain. Think of your own reactions to these factors.

When we study behaviour we ask ourselves questions like 'What stimulus caused that response?' 'What purpose did the response serve?' 'How did this response come about, was it inherited or was it the result of learning?'

When we watch a thrush hammering a snail against a stone, we recognise at once that this kind of behaviour is useful to the bird because after cracking the shell it can peck out the edible part. When a fly takes off as soon as your hand approaches, it is immediately placed out of danger. When a lion approaches a herd of buffalo they all bunch together and face it. These examples show that animals tend to behave in ways which are favourable to their existence, i.e. their behaviour is an aid to their survival.

In order to survive, an animal has to carry out various vital activities such as feeding, reproducing and protecting itself; consequently, attempts have been made to classify behaviour patterns according to the functions they serve. Nine categories have been described, although the simpler animals do not carry out all of them:

1. Eating behaviour
This varies greatly in different species; dogs gulp their food; other mammals crush their food before it is swallowed. Bees suck up their food, flies lick it up. Anteaters use their tongues, sharks use their teeth; we use a knife and fork.

2. Shelter-seeking behaviour
This is the tendency to find the best conditions and to avoid dangerous and harmful ones. For example, birds roost in places where they find protection, starfish and crabs hide under boulders and shrimps bury themselves in sand. We like to take the chair nearest the fire, or put up an umbrella if it rains. Have you ever watched a flock of sheep or a herd of cows in driving snow or rain? Which way do they move? From your answer you should be able to work out why sheep are more liable to be buried in snow-drifts than cattle.

3. Aggressive behaviour
This includes fighting and **competing for dominance**. Stags show this at the mating season when they clash their antlers with those of rival stags and try to force them off their territory. Many birds show it when they defend their territories by displaying (e.g. robins puff out their red breasts) or flying at intruders and chasing them away. We show it when we compete in various sports, lose our tempers and fight wars.

4. Sexual behaviour
Many animals have elaborate courtship patterns of behaviour. The peacock erects its strikingly-coloured tail, bower birds decorate their nests with flowers, spiders and scorpions perform dances and newts vibrate their tails. We put on our best clothes and give flowers or boxes of chocolates.

5. Care-giving behaviour
Many animals protect their young in various ways. Birds build nests, incubate the eggs and feed the young; mammals suckle their young, protect them from predators and bring them food. We build houses for our families, protect our children, find food for them and teach them.

6. Care-soliciting behaviour
This is asking for something which is wanted or needed. Young birds show this type of behaviour when their parents return to the nest to feed them. They stretch up their

Fig. 10:1 Nesting cape cormorants: each has a territory round its nest roughly determined by the pecking distance between neighbours.

necks, open their mouths wide and display their gape which is often brightly coloured and so attracts attention; at the same time they often make loud noises. Similarly, a human baby cries to attract its mother's attention and older children pester their parents to give them sweets or ice creams.

7. Eliminative behaviour

This relates to the elimination of waste products; such behaviour is seen when cats dig holes for their faeces and parent birds remove the faecal pellets of nestlings in their bills, thus preventing the fouling of the nest.

8. Contagious behaviour

This is sometimes described as mutual mimicking; it occurs when two or more of a species do the same thing. This includes the flocking behaviour of birds, shoaling in fish and herding in cattle. We notice it when one person yawns and others do the same. Watch a group of two or three people talking together and see how unconsciously they imitate each other by such movements as folding the arms. It becomes more obvious among spectators at football matches.

9. Exploratory behaviour

This is the tendency to explore an unfamiliar habitat. When a mouse is put in a cage it will quickly explore every part of it. A dog or cat will do the same if put in a room it has not been in before. It is very marked in us—we show great curiosity. When we go to a new place for a holiday we want to investigate the whole area. Our desire to travel, climb mountains and do research are other examples.

Social behaviour is the term used to describe any of the above categories when two or more individuals of the same species are involved. This becomes very marked when many members of a species live together in large social groups or societies such as the ants, bees and wasps—not forgetting ourselves.

Keeping this classification in mind, you would find it interesting to analyse your own behaviour during a day; try to decide into which category each main behaviour pattern comes.

We will now study one aspect in more detail
—shelter-seeking behaviour.

Shelter-seeking behaviour

In a forest the animals are distributed throughout the habitat. An experienced naturalist
would look in a definite place for a particular
species. He would look underneath leaves of a
certain kind of tree for one, under the bark for
another, in the soil for something else, under an
old log, in the leaf litter, amongst the mosses
and lichens, and so on. One of the reasons
why animals are found in certain places is
because they exhibit shelter-seeking behaviour. If you find a caterpillar on the underside of the leaf and put it on the upper surface,
it will crawl back again; lift up a log or a stone
and many creatures which are at first visible
will soon disappear from view. This is because
they react to such factors as light and move
until they are in darkness again. There are
many other factors which influence their directional movements, such as humidity, wind,
temperature and scent. Usually, an animal's
behaviour is determined by more than one of
these factors acting together. If you come
across a fox or badger hole in a bank or wood,
you may see flies constantly going in and out of
the entrance; this is a good sign that an
animal is living there. They fly *in* because they
are attracted to the smell of the animal, but in
doing so they get into the dark; this causes
them to fly *out* again towards the light, hence
they shuttle to and fro!

We can now investigate for ourselves why an
animal such as a woodlouse is to be found in
leaf litter, under stones or logs or under loose
bark. These are all damp, dark places, so we
might investigate whether this is due to either
or both of these factors.

1. The effect of humidity:

Use the apparatus shown in Fig. 10:2, or devise
something similar: it is known as a choice
chamber. The humidity of the two chambers
can be altered by putting water under one
chamber and calcium chloride or silica gel
under the other (these chemicals absorb water).

Place a strip of dry (blue) cobalt chloride
paper in each and put the sheet of glass on top.
When the strip in the chamber above the
water turns pink you will know that a distinct
difference in humidity within the two chambers
has been reached. Now quickly introduce an
equal number of woodlice into each chamber—
about 5 will do—by slipping the glass to one
side just enough to put them in; in this way
very little mixing with the outside air will
occur. In order to eliminate the possible effect
of light on the behaviour of the woodlice, place
the chambers in such a position that they
receive the same amount, i.e. they are at the
same distance from the main light source.

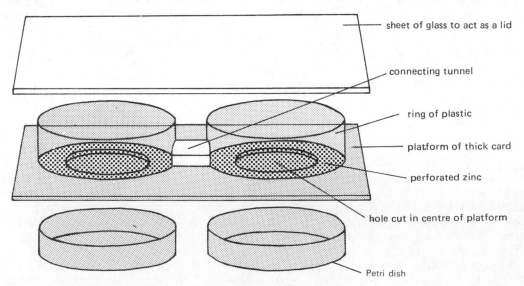

sheet of glass to act as a lid

connecting tunnel

ring of plastic

platform of thick card

perforated zinc

hole cut in centre of platform

Petri dish

Fig. 10:2 Choice chamber. Component parts have been separated for clarity.

Every minute record the number of woodlice in each chamber over a period of 10 minutes. Also try to estimate whether the woodlice are more active in one than in the other. Think how you could do this accurately.

What conclusions can you draw from this experiment?

2. The effect of light:

Carry out a similar type of experiment to show the effect of light and darkness, by shining a bench lamp on one chamber and covering the other with a black cloth. How would you eliminate the heating effect of the lamp? Keep the humidity the same in each and quickly count the number of woodlice in each chamber at minute intervals as before. What do you conclude from your results?

In terms of activity, it has been found that animals often move faster when experiencing adverse conditions. This enables them to find favourable conditions more quickly. Was this true of the woodlice?

You have now investigated their behaviour in relation to two factors, light and humidity. There could, of course, be other influential factors as well, and some might be more important than others in producing a response in a given situation. One of these other factors is contact. Earwigs, millipedes, centipedes and woodlice are often found in similar situations and react to light and humidity changes in a similar way, but once in a dark and humid place they often stop at a place where much of their body surface is in contact with something firm. This is why you often find them in cracks or tunnels. Woodlice and earwigs are often found clustered together; could this be due to contact or because of the greater humidity resulting? All sorts of questions come into your mind when you try to analyse behaviour patterns! What factors do you think cause trout to face upstream, salmon to find their way back to a particular river and fleas and mosquitoes to find a victim?

Sign stimuli

Not all rigid behaviour patterns are responses to such factors as light, temperature or humidity. Some are the result of sign stimuli. For example, a young herring-gull chick will peck at the red spot on its parent's bill, causing the mother to regurgitate food from its crop. If you present to a chick a gull's head with a red spot on it, cut out roughly in cardboard, it will peck at the artificial spot just the same (Fig. 10:3). Experiments have been done to discover the chick's reactions when the colour of the spot and the shape of the head and beak are varied, and to find out which colours and shapes act as the best sign stimuli.

David Lack, a pioneer in the field of bird behaviour, described in his book, *The Life of the Robin*, how he investigated the territorial behaviour of this species. He had watched robins displaying and chasing other robins off their territories and he wondered how a robin would react if he placed a stuffed one on its territory. So he wired a stuffed specimen to a branch near a robin's nest which contained young. When the parent returned and saw the stuffed one, it flew towards it and displayed its red breast—getting no response, it attacked it fiercely and pecked it to pieces! Was it the red breast that caused the robin to attack? To find out, Lack repeated the experiment using another stuffed robin, but this time he covered over the red feathers with brown paint. The live bird took no notice of it! Surprisingly, he

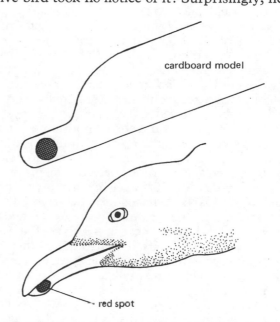

Fig. 10:3 The red spot on a gull's bill acts as a sign stimulus to a young chick which then pecks at it to obtain food. Cardboard models of various shapes have the same effect.

117

found that even an isolated bunch of red feathers would cause a robin to attack. So it was the red colour that acted as the sign stimulus which initiated the attack.

Do you see any significance in the fact that *young* robins have no red breast?

Sticklebacks also use sign stimuli. At the breeding season the male develops a red underside which acts as a deterrent to other males which swim near. If the intruder persists it will be attacked and driven off.

If you have in an aquarium a male which is in breeding condition, you can test this reaction for yourself by putting a mirror in the water. What happens when it sees its own image in the mirror? You could also make a simple model of a fish and paint its underside red and dangle it on a wire in front of the live male and note its reactions. If sticklebacks in breeding condition are unobtainable you could investigate in a similar way the reactions of such tropical fish as Siamese fighters.

Niko Tinbergen, a leading authority on animal behaviour, and other research workers were helped by experiments such as these to discover that the whole process of reproduction in the stickleback could be seen as a series of reactions to sign stimuli, each one triggering off the next. The main steps are as follows:

Once its territory is secure the male stickleback constructs a nest. It makes a shallow pit in the sand, collects small pieces of weed and glues them together by means of a sticky secretion. It then forces its way through this nest, so moulding it into the shape of a tunnel. If a female swims near, her bulging abdomen (due to the eggs inside) stimulates the male to perform a zig-zag dance round her, thus displaying his red underside. If the female is ready to lay she responds by curving her head and tail upwards. This behaviour stimulates the male to swim to the nest, causing her to follow. The male then prods the entrance with his snout, causing the female to push past him into the nest. The sight of the female's tail projecting from the nest causes the male to prod at it with his snout; this causes her to lay. When she swims off, he enters and fertilizes the eggs. The male then guards them persistently until they hatch, fanning the water with his tail; this helps to keep the eggs aerated. After they

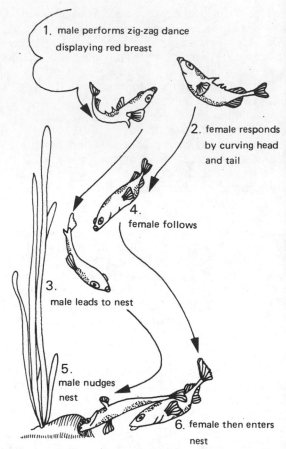

1. male performs zig-zag dance displaying red breast

2. female responds by curving head and tail

3. male leads to nest

4. female follows

5. male nudges nest

6. female then enters nest

Fig. 10:4 The succession of sign stimuli used during stickleback courtship.

hatch, the male guards the young and if they stray sucks them into his mouth and spits them back into the nest (Fig. 10:4).

Instinctive behaviour

Behaviour patterns such as these are not learnt but, in some way not understood, are passed from one generation to the next; they are thus said to be **innate**. Such reactions are characteristic of what is often called **instinctive behaviour**. However, there is no animal which shows no variation in its behaviour; all appear to be able to learn from experience to some extent. Spiders are good animals to study in this respect.

Find a specimen of the common house spider *Tegenaria sp.* It is the large one which you sometimes find in baths or sinks. You can

Fig. 10:5 a) Male stickleback on nest. b) Chick pecking at the red spot on a cardboard model of a gull's head. c) Social behaviour: baboons living together in a mixed group. d) Starlings coming in to roost: an example of both contagious (flocking) and shelter-seeking (roosting) behaviour. e) Shoaling of fish: an example of contagious behaviour. f) Ducklings following the hen which brought them up: an example of imprinting.

119

often find these spiders in sheds, cellars or among logs. Put one in a cardboard box, using a piece of glass as a lid. Put in one corner a small box with one side removed to provide a dark place for the spider to live in. After a day or so it will have made a sheet web all over the box. Now remove the glass, take a tuning fork and touch the web with the vibrating fork. The spider should rush out and attack it as if it were a fly. This is an innate response to vibrations of the web. Finding that the fork is not a victim, it will retire to its home. Repeat the procedure at two minute intervals. How long does it take the spider to learn not to react?

In these examples we have shown how animals respond to a particular stimulus in a definite manner, but in many instances animals are subjected to more than one stimulus at the same time and these may be conflicting. For example, if food is placed on the lawn and a cat is nearby, a hungry bird on seeing both the food and the cat is faced with conflicting stimuli —to approach and eat the food and to fly away from the cat. Clearly, it cannot respond to both stimuli at the same time. In this instance the fear reaction usually suppresses the response to food. So there must be some form of internal control which determines which pattern of behaviour is followed. However, in many other instances an animal may be stimulated strongly in conflicting ways but neither behaviour pattern becomes dominant; instead the animal does something quite irrelevant to the situation. This is called a **displacement activity**. For example, a bird when confronted with a rival may be equally stimulated to fight or fly away—in the event it does neither, but starts to preen its feathers. A cat under similar circumstances may start to lick itself. We show similar displacement activities when we have conflicting emotions—we scratch our heads, tap our fingers on the desk, or adjust our clothes.

Rigid and adaptable behaviour

The types of behaviour we have considered so far have been rigid, although each is capable of being modified to a very limited extent as a result of learning. Some animals, by contrast, rely much less on rigid, innate responses; they learn rapidly, and their behaviour is governed largely by the experiences they have built up. We can describe this kind of behaviour as **adaptable**; it is particularly characteristic of mammals, and especially man.

An animal which relies on rigid, innate patterns of behaviour has a set of ready-made answers to all the usual problems it is likely to meet, and this is true as much for the young as for the adult. This is of great survival value when conditions are normal, but when the animal is faced with unusual circumstances it may be disastrous.

An animal which relies mainly on learning is at a disadvantage when young, as it takes time to gain experience. However, adaptable behaviour is usually associated with a high degree of parental care, and this helps the young to survive until they have gained the necessary experience. Once the early stages are passed, adaptable behaviour allows an animal to cope better with unusual and difficult circumstances.

Innate or learnt?

Many animals, such as the majority of insects, never see their parents, as the latter are dead by the time they hatch and so they have no chance to learn from them; thus most of their behaviour is innate. With vertebrates there is a greater opportunity for learning. It is not always easy to know if a particular type of be-

Fig. 10:6 Common house spider (*Tegenaria sp.*) × 3.

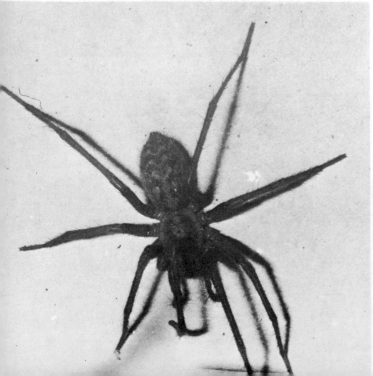

120

haviour is innate or learnt; for example, when a bird sings does it inherit the song or learn it? In an experiment to investigate this, various species of birds were reared in incubators and isolated completely from other members of their species. When they became mature and started to sing, the songs of some species were exactly the same, proving that these songs were innate; but in others the song was a very poor replica of the normal. However, in the latter case if the young birds were allowed to hear an adult sing just once at about hatching time, it was found that they sang the complete song when they became adult. This is an example of a type of learning called **imprinting**, which we will consider next.

Learning

We can distinguish several different types of learning:

1. Imprinting and early learning
Imprinting is extremely quick learning and is characteristic of some birds and mammals during a brief period when they are extremely young. **Konrad Lorenz**, an Austrian scientist famous for his studies of animal behaviour, discovered that newly-hatched ducklings accepted as their mother the first large moving object that they saw in the vicinity of the nest. Normally this would, of course, be their mother, but by raising the eggs in an incubator, Lorenz made sure that he was the first object to be seen when they hatched. Consequently, they followed him about everywhere, just as if he was the mother duck!

In the same way many mammals, if brought up artificially on the bottle, will treat their owner as if he or she were their mother. The nursery rhyme about Mary and the little lamb was biologically accurate! The period of imprinting often coincides with the time when the eyes are first becoming functional.

The learning which occurs early on in the life of a mammal is undoubtedly very important in the development of the personality of the individual concerned. Baby rhesus monkeys, for example, when brought up artificially without access to their mothers, were found to be incapable of normal relationships with monkeys of either sex when they grew up. If, when small, they were given something furry or soft to hug, they would run to this when frightened, as if it were their mother. By having this substitute mother, they partly overcame the handicap shown when they were completely deprived of their own mothers.

Does this kind of behaviour apply to humans too? Although not easy to prove because of the difficulty of making adequate controlled experiments, more and more evidence is accumulating that what happens to a child between the ages of a few months and two or three years is of very great importance. A young child needs such basic comforts as food, warmth and bodily contact from its mother or foster mother. If the child receives and can rely on these comforts when required, and is not left to cry for long periods on its own, it feels more secure and is more likely to grow up to be more responsive, more eager to learn and to have easier relationships with others. If these normal comforts are denied to a young child he is more likely to grow up into a 'difficult' personality. In other words, what a child needs most of all is to be loved and to have the security of a home and happy family background. But being loved is not the same thing as being given everything that is wanted. Wise and consistent discipline is an essential ingredient of love, as only this will lead to self-discipline later on.

2. Habituation
This is brought about when a natural response is lost. Young animals are frightened by loud noises, but later they learn to discriminate and only react to those they have learnt are significant. If we live by a railway or busy road we soon become habituated to the noise and become unconscious of it, but a visitor from the country might find the noise most disturbing. Nocturnal animals have been filmed successfully, using artificial light which normally would have frightened them, by gradually allowing them to become habituated to increasingly intense light over a long period beforehand.

3. Conditioning
The Russian scientist **Pavlov** (1849–1936) performed many famous experiments on the learning behaviour of dogs; he used the simple reflex, whereby saliva is secreted when food is taken into the mouth, as a basis for his work. This reflex is innate, but he demonstrated

Fig. 10:7 Exploratory behaviour: stoat on tree stump.

10:9 Eating behaviour: young lions on a dead buffalo.

Fig. 10:10 Aggressive behaviour: red deer stags fighting.

Fig. 10:8 Care-soliciting behaviour: young marsh warblers stretch necks and display their coloured throats.

Fig. 10:11 Shelter-seeking behaviour: snails clustered in a leaf of aloe.

122

Fig. 10:12 Defensive behaviour: when attacked, this grasshopper suddenly displays its red underwings, momentarily surprising the aggressor and so gaining time to fly off.

Fig. 10:15 Sexual behaviour: peacock displaying before hen.

Fig. 10.13 Territorial behaviour: red deer stag roaring, with group of hinds.

Fig. 10:14 Care-giving behaviour: chimpanzee mother with her baby.

Fig. 10:16 Intelligent behaviour: a chimpanzee piles boxes on top of each other to obtain food hung out of reach.

that if he rang a bell every time he fed the dog there would come a time when, by ringing the bell without giving any food, saliva would be secreted just the same. In this way the original stimulus is replaced by a different one, but the response remains the same. This is the essence of a **conditioned** reflex.

4. Trial-and-error behaviour

This type of learning may be looked upon as an extension of the last. It occurs when an association is built up between a certain action and a reward or punishment. For example, young birds rapidly learn to avoid eating insects which are yellow and black once they have taken such a specimen which is either bad tasting or can sting.

Pets brought up in the home, or animals trained in a circus, associate reward or punishment with certain actions and so learn to behave in a particular way. In this manner they are taught to do most complicated actions. You will have noticed how the circus trainer will give a reward whenever the animal carries out the action successfully, but withholds the reward if it does not. The same applies to us; small children are helped to behave in a particular way by praising them or rewarding them when they do well. If we ride a bicycle recklessly and consequently have an accident, we learn to ride more carefully as a result of the experience. The expression 'learning the hard way' embraces many examples of trial-and-error behaviour. Can you think of other examples?

5. Insight learning

This is the highest type of learning and is best shown in apes and man. Insight learning involves reasoning—the situation is considered in the light of past experience and the problem is dealt with in a particular way. For example, a chimpanzee was put in a cage with a bunch of bananas hung out of reach, and some boxes were left on the floor. The ape had never been in this situation before. The chimpanzee looked at the bananas and considered the problem, then it piled the boxes on top of each other, climbed up and reached the bananas. This is an example of **intelligence**, which may be defined as the ability to organise behaviour in the light of experience.

Insight learning is of great survival value because it enables an animal to adapt to changing circumstances; it is developed to a very high degree in man and is the basis of many of his great achievements.

Behaviour and social organisation

Some animals have an increased chance of survival if they live together in societies. Consequently any aspect of behaviour which helps to maintain the organisation of the group is important. We saw in Book 1 p. 78 that the social behaviour of bees is remarkably complex, the behaviour of each individual being related to the well-being of the whole colony. In this society the complex role played by each bee is almost entirely innate, for its life span is too short for it to learn all that is needed and its brain is not complex enough.

As we have seen, in many of the higher vertebrates learning about the environment through processes of habituation and trial-and-error behaviour constitutes an important part of an individual's development. In the case of social vertebrates such as jackdaws, wolves, monkeys and humans, the environment includes other members of the same species. Because individuals in such societies may vary in age, size and fitness, there is often a hierarchy, or **'peck order'**. In birds such as hens or pigeons it is linear; this means that there is a dominant bird A which can peck any other without being pecked, B which can peck all but A, C all but A and B, and so on. The unfortunate one at the bottom of the order may be persistently pecked and fail to obtain enough food in consequence. Such a system ensures that in times of hardship the strongest members of the group are the first to have access to food. An individual soon learns his position in such a hierarchy through fighting or trials of strength, but once determined the situation is usually accepted and so there is less aggression.

In mammals the relationships are seldom linear, but there are often dominant males as in cattle and deer herds, and dominant females in sheep. The dominant lion in a pride has the privilege of eating first and obtaining the best pieces; the juveniles have to wait their turn. We notice similar signs of a hierarchy in human society, for example in a family, class, school or team.

In highly developed animal societies not only do such mechanisms exist to restrict aggression, but care-giving behaviour between individuals makes a positive contribution to the harmony of the group. This is developed to its highest levels in humans with the concepts of unselfishness and love. It would seem that the greatest problem of human society today is learning how to live together; without the application of these higher aspects of care-giving behaviour civilization will have no future, as aggressive and anti-social behaviour will ultimately lead to the break-up of society.

11

Living together

Ecology

In the next two chapters we shall be studying the principles of ecology. This is a branch of biology which deals with the relationships between living organisms and their environment.

In Book 1, when studying the vertebrates, we saw how fish were adapted to an aquatic environment, how birds were adapted for life in the air and how mammals fitted into a variety of contrasting habitats. But fish, birds and mammals, and any other group for that matter, cannot live in isolation; they are dependent on many other kinds of organisms for their vital needs. So in every habitat there is a great variety of plants and animals all living together and, to a varying extent, dependent upon each other. Groups of animals and plants which live together in a particular habitat are called **communities**.

The size of the community varies with the size of the habitat. We talk of terrestrial communities of organisms, such as those living in deserts, forests and in the soil; aquatic communities, including those in a puddle, a lake or an ocean; there are communities to be found in a pile of dung, in the intestine of a living animal or in the rotting body of a mouse. You will be able to think of many others.

Each community is characteristic of the particular habitat in which it is living; thus the flora and fauna of a rain puddle are very different from that of a rock pool, and the life of a pine forest is very different from that of an oak forest. This is because the organisms com-

prising a community are specially adapted to live under these particular conditions, but in other habitats, like a 'fish out of water', they would probably not survive.

Because of this close relationship between community and habitat they are always studied together, and we refer to a community and its associated habitat as an **ecosystem**. Ecology is the study of ecosystems.

Ecology is a subject of very great importance to all of us, because the principles involved can lead us to an understanding of how man himself can live in balance with all other living things, utilising in the best way the limited resources at his disposal without destroying the environment. Man's survival depends basically upon whether he can apply the principles of ecology to himself.

Ecosystems

All life on our planet is confined to a thin envelope consisting of the atmosphere, oceans and earth's crust. This region—the world of living things—is termed the **biosphere**.

Within the biosphere there are a number of major ecosystems, the terrestrial ones being determined largely by the variations in climatic conditions between the Poles and Equator (Fig. 11:1). In a similar way, if you climb a mountain such as Kilimanjaro (5811m) in Equatorial Africa, you quickly go through a comparable system of ecosystems, starting with tropical rain forest at the base and ending with perpetual snow and ice at the summit (Fig. 11:2).

The main climatic influences which determine these ecosystems are rainfall, temperature and the availability of light from the sun. For instance, forests are usually associated with high rainfall, but the type is influenced by temperature and light; the same applies to deserts which occur in regions where rainfall is extremely low.

Figures 11:4 to 11:8 illustrate the typical appearance of some of these ecosystems.

1. Tundra

This region, which adjoins the permanent ice of the polar region, is devoid of all trees, but stunted shrubs such as birch and sallow occur in its more southern parts. The ground flora include many lichens, mosses and sedges. The soil is frozen for most of the year, but the top

126

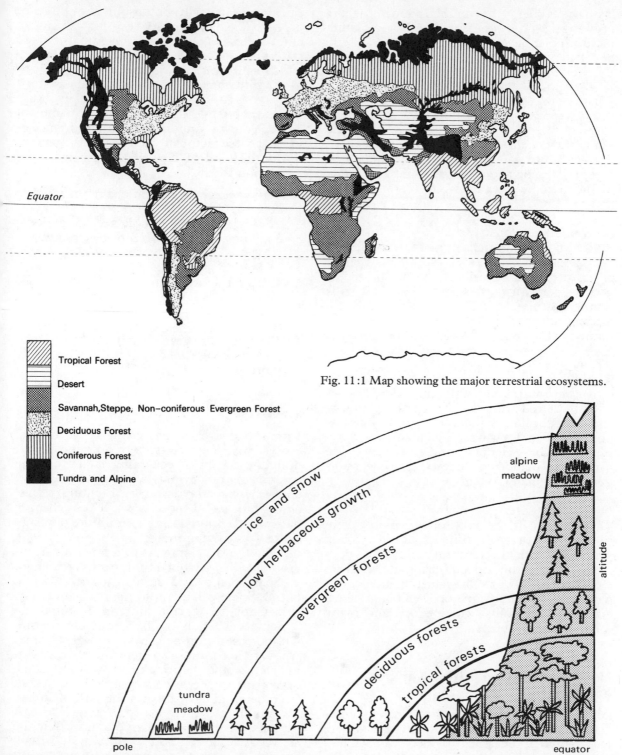

Equator

Tropical Forest

Desert

Savannah, Steppe, Non-coniferous Evergreen Forest

Deciduous Forest

Coniferous Forest

Tundra and Alpine

Fig. 11:1 Map showing the major terrestrial ecosystems.

ice and snow

low herbaceous growth

evergreen forests

deciduous forests

tropical forests

alpine meadow

altitude

tundra meadow

pole

equator

Fig. 11:2 Diagram showing how mountain climatic regions can be compared with the horizontal climatic regions of the earth.

layer melts during the summer, allowing a short growing season of about two months. The fauna include polar bear, Arctic fox, reindeer, Arctic hare, lemming, snowy owl and ptarmigan.

2. Coniferous forest
This region occurs south of the tundra. Here the winters are not so long and the greater summer warmth allows trees to develop. In the north they occur only in sheltered places, but further south as extensive forests dominated by spruce, pine and other conifers. The fauna include lynx, wolverine, wolf, elk, red squirrel and grouse.

3. Deciduous forest
This region lies to the south of the coniferous belt. Most of Britain lies within this region. It has been greatly reduced in size and modified by man's activities, but what remains is dominated by such trees as oak, beech, birch, ash and alder with many shrubs and herbaceous plants. The fauna include fox, badger, red deer, roe deer, mole, mice and voles.

4. Savannah
This is a tropical region dominated by grasses with scattered trees and fire-resisting thorny shrubs. The fauna include a great diversity of grazers and browsers such as antelopes, buffalo, zebra, elephant and rhinoceros; the carnivores include lion, cheetah, hyaena, mongooses and many rodents.

5. Tropical forest
This occurs in the equatorial region where rainfall is heavy. It consists of lush forest vegetation with tall trees and woody vines with stems that climb up and hang down from trees, called **lianas**. The fauna include chimpanzee, monkeys, okapi, forest elephant, small antelopes, hornbills, woodpeckers and many other species of birds.

Natural variation within a major ecosystem

Although each major ecosystem has its characteristic climatic conditions and typical flora and fauna, it becomes divided into smaller ecosystems which differ markedly from the basic type because of variation in local conditions and especially because of the influence of man. We have already seen how change in altitude influences vegetation and therefore the fauna too, but there are many other factors which cause variation in the ecosystem, such as the influence of the sea and inland waters, the wetness of the soil, the degree of exposure to wind and the geology of the area, to name just a few. You will understand the importance of some of these factors if you consider in general terms the kind of changes occurring in the vegetation when:

1. A river, such as the Nile, runs through a desert region.
2. Low-lying land is subjected to periodic flooding a) by salt water, b) by fresh water.
3. A region is subjected to very high winds, such as on an island.
4. The land slopes steeply as in a mountainous region.
5. There is a change from a sandy to a calcareous soil.
6. There is a change in the amount of exposure to the air due to tidal effects on a rocky shore.

The effect of man on ecosystems

In many areas man has changed the natural ecosystems very greatly by damming rivers, draining marshes, reclaiming land from the sea, cutting down forests, ploughing up land and growing crops, and by building towns, cities, canals and motorways. These changes have greatly altered the communities of plants and animals living there.

Take the development of a large town, for example. There will be three kinds of change: 1. Some plants and animal species will die out. 2. Some will adapt to the new conditions sufficiently to survive in reduced numbers. 3. Some will benefit by the new conditions and will increase in numbers. Many of these changes will vary according to where the town is but, as you would expect there will be some organisms in the last category which benefit from being in close association with man and his buildings, such as sparrows, starlings, pigeons, rats, mice and many plants we describe as weeds. Can you add to this list and think out which plants and animals in your area would come into the first two categories?

Interactions within the ecosystem

These interactions are complex, but the general principles are illustrated in Fig. 11:3.
1. The physical environment is mainly determined by the interaction of three kinds of factors: a) **climatic**, e.g. rainfall, temperature, light; b) **edaphic** (soil), e.g. type of soil, water content of soil; c) **physiographic**, e.g. altitude, slope.
2. The physical environment interacts with the flora and largely determines its composition. We have already mentioned that rainfall and temperature are particularly important in determining the presence of forest or desert, but this is a two-way relationship as the flora itself can affect the physical environment; for example, the presence of forest may increase the rainfall because of the high rate of transpiration from the leaves.
3. The physical environment also affects the fauna. In the Arctic, for instance, the intense cold restricts the number of terrestrial species to warm-blooded birds and mammals. On the sea-shore the distribution of a number of species is affected by the strength of the waves. The fauna can also affect the physical environment; for instance, when hippopotamuses are too numerous in parts of Africa, their grazing is so severe that much soil erosion takes place during tropical rainstorms. You will also remember how the activities of earthworms can alter the nature of the soil (Book 1 p. 41).
4. The flora interacts with the fauna in a great many ways. All animals are dependent upon green plants for food and oxygen, either directly or indirectly, and the species of plants which are present will determine the kinds of animals which can live there. As well as using it for food the fauna will also affect the flora through factors such as trampling, manuring and pollination. These relationships will be treated more fully later.
5. There are also interactions between the plants themselves. There will be competition between species for light and mineral salts; some species will act as parasites on other plants, while species such as those which form lichens will live together in symbiosis (p.135).

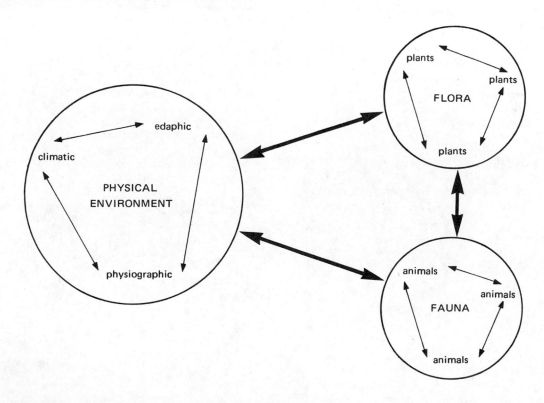

Fig. 11:3 Diagram illustrating the various interactions within the ecosystem.

Fig. 11:4 (above) Coniferous forest: Sweden.

Fig. 11:5 (below) Savannah (with gazelles): Tanzania.

Fig. 11:6 Tropical rain forest: Guyana.

Fig. 11:7 Deciduous forest: England.

Fig. 11:8 Tundra (with caribou): Alaska.

6. Finally, there are many interactions between the animals present: between predator and prey, parasite and host and through many aspects of competition both between different species and between members of the same species.

The last three categories listed above are described as **biotic** factors as they concern the interactions of living organisms within the ecosystem.

Food relationships

All living organisms need energy. The ultimate source of all energy is sunlight and only one type of organism can harness this energy for its own use—the green plant. This can be done because chlorophyll absorbs the light energy which is used to build up sugars in photosynthesis. In addition, the green plant, with the help of nutrients from the soil, can build up more complex substances such as proteins.

All these materials synthesised by green plants become the *only* source of food for non-green plants and animals, directly or indirectly. It is this food that becomes the source of energy for all these organisms and which provides material for their growth. Thus green plants in any ecosystem are called the **producers** and the remaining organisms which are all dependent upon them are the **consumers**.

Animals which feed directly on plants are called **first-order consumers** (herbivores); elephants, antelopes, rabbits, snails, caterpillars and aphids all come into this category. Within their bodies the food is digested, absorbed and built up into their own tissues, which then become the food of the animals which feed on them; these are the **second-order consumers** (carnivores). They range in size from lions which feed on antelopes, foxes which prey on rabbits, warblers which eat caterpillars to ladybirds which feed on aphids. **Third-order consumers** may be present in some ecosystems; they feed on the smaller of these carnivores, e.g. the hawks which prey on warblers.

These relationships based on feeding habits are called **food chains**, e.g.

Oak tree → aphid → ladybird → warbler → hawk.

Grass → antelope → lion.

But these links are never as simple or rigid as the word 'chain' suggests. For example, aphids are eaten by many insectivorous birds in addition to warblers, and also by ladybirds and other insects; hawks, on the other hand, prey upon a considerable variety of birds and

131

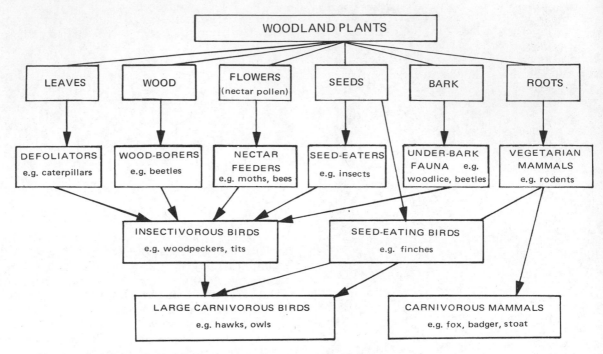

Fig. 11:9 A food web characteristic of a woodland ecosystem (simplified).

small mammals. So the term **food web** is often a better one to use when being precise, as it suggests a far greater number of possible links and reflects the fact that the whole community is a complex inter-connected unit.

Figure 11:9 is a much simplified diagram to show some of the feeding relationships amongst organisms living in deciduous woodland. You will see from the diagram that animals fit into special positions within the food web; each is described as its **niche**. For instance, there is a niche for insects such as aphids which suck up the juices of leaves by means of a proboscis; another niche for insects such as caterpillars which have strong jaws for biting off pieces of leaf; and a niche for relatively large animals such as deer which browse on the vegetation. All these animals feed on leaves but they differ both in size and in the manner in which they feed. So the term 'niche' denotes not only the animal's position in the food web and what it eats, but also its mode of life. Just as a habitat is the place where an animal lives, so a niche describes its occupation—the way it 'goes about its business and earns its living'.

Food chains involving scavengers

When plants and animals die or produce waste substances, their products form the food of other species which are known as **scavengers**. Thus dead plant material, **humus**, may be eaten by earthworms or woodlice; dead animal matter, **carrion**, by maggots, beetle larvae or in some ecosystems, vultures; and the dung of animals by flies or dung beetles. These scavengers in turn become the prey of carnivorous birds and mammals, which again may be eaten by still larger carnivores.

Much dead material is not eaten but decays, due to the activity of fungi and bacteria—these are described as **decomposers**; they play an important part in returning valuable material to the soil in a form which can be utilised by green plants once more (p.135).

Food chains involving parasites

We saw an example of such a chain when studying the large white butterfly (Book 1); how their caterpillars were parasitised by the ichneumon *Apanteles*, which in turn was parasitised by another ichneumon *Hemiteles* and these in their turn by a Chalcid wasp.

Many food chains involving parasites are more simple. For example, the ticks which feed on the blood of buffaloes are themselves eaten by oxpecker birds which may in turn be eaten by a predatory bird such as a hawk.

Size and numbers in relation to food chains

At each link in a food chain, from the first-order consumers to the large carnivores, there is normally an increase in size, but a decrease in number. For example, in a wood, the aphids are very small and occur in astronomical numbers, the ladybirds which feed on them are distinctly larger and not so numerous, the insectivorous birds which feed on the ladybirds are larger still and are only present in small numbers, and there may only be a single pair of hawks of much larger size than the insectivorous birds on which they prey. This relationship is best shown as a pyramid (Fig. 11:10). An exception to this is the parasitic food chain where the parasite is always smaller than its host. Can you think why this must be so?

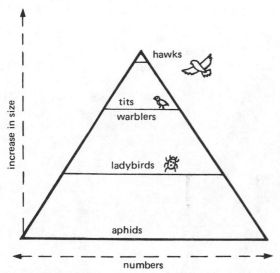

Fig. 11:10 Diagram to illustrate the principle of the pyramid of numbers.

The weight pyramid

A more accurate idea of food relationships may be obtained if this pyramid of numbers is converted into a pyramid of weights. This would indicate the weight of plant matter which is used by the aphids to produce the weight of the aphid population, the total weight of the ladybird population that could be supported by the aphids and so on through the chain.

The weight pyramid shows that animals are relatively inefficient in converting food into body tissues, the remainder of the food being undigested and passing out as waste, or broken down in respiration to supply energy for such activities as feeding. Many animals convert no more than 10% of their food into body tissues, some herbivores even less. Let us take an example of a food chain which has been worked out in some detail—one in which we are involved when we eat fish. In this chain the plant plankton in the surface waters of the sea trap energy from sunlight and are the food producers; the animal plankton feed on the microscopic plants and the fish in turn feed on the animal plankton; we are at the end of chain when we eat the fish.

Plant plankton → animal plankton → fish → man

In this particular food chain roughly 90% of the food is lost at each step, so it follows that it would take 1000kg of plant plankton to produce 100kg of animal plankton to form 10kg of fish to produce 1kg of human tissues, with a corresponding loss of the original plant potential energy that came from the sun. So the nearer an animal species is to the original plant source in a food chain the greater the amount of energy is available to the population of that species. In other words, the fewer the steps in the food chain, the more energy there will be for the species at the top. This principle is of very great importance to man with his problem of feeding the world's expanding population.

You should now be able to work out why it is so wasteful to turn fish into fish meal for cattle and then eat the beef, instead of eating the fish direct; why vegetable food is cheaper than meat and fish; why it is more economic to breed vegetarian fish, e.g. carp, for human consumption, rather than carnivorous ones, e.g. trout.

Recycling materials in an ecosystem

This constant dissipation of potential energy as it flows through an ecosystem is replaced by energy from the sun through photosynthesis. But what of the materials such as oxygen, carbon dioxide, water and mineral salts which are being used by organisms; why does not the supply run out? After all, living organisms have been using up these substances for millions and millions of years and there is no source of replacement for them outside the earth's atmosphere. The answer is that all these essential substances are somehow replaced and used over and over again; that is, they are **recycled**. Let us first consider why the amount of oxygen and carbon dioxide in the air remains so remarkably constant.

What do you think would happen if plants were grown in damp soil in a large glass bottle which was sealed from the air? Would the plants remain alive? Would either the oxygen, carbon dioxide or water be used up after a time, or would they be recycled? You could find out by setting up this miniature ecosystem as a class demonstration:

Place 5kg of sterilised soil in a glass carboy of about 25 litre capacity. Add enough water to make the soil uniformly damp, but not waterlogged. Take four cuttings (about 15cm long) of *Tradescantia*, drop them into the carboy and push their cut ends into the soil

with a stick (this species roots easily). Close the carboy with a rubber bung to keep out the air and place the apparatus in good light, but not direct sunlight. Write the date on the carboy at the start of the experiment. Note whether the plants grow and, if possible, how long they live.

'Bottled gardens' of this kind sometimes flourish for many months and sometimes years. Attempt to explain how recycling is going on in a 'bottled garden' by considering the processes of photosynthesis, respiration, water absorption and transpiration. Why is it important not to put the apparatus in full sunlight?

We will now study in more detail some of these recycling processes, considering the whole world as an ecosystem.

The recycling of water

Over a long period of time there is a rough balance between the water which is being precipitated as rain, dew, hail or snow from the atmosphere and that which is evaporated from the surface of the land, oceans and other bodies of standing water. A variation of this cycle

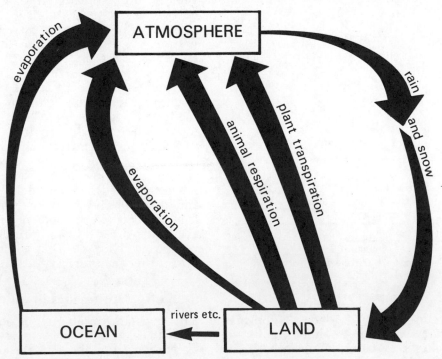

Fig. 11:11 Diagram showing the recycling of water.

occurs when water is taken in by plants and animals and passed on to others through food webs, or given out once more in transpiration, excretion, respiration and other processes. But again there is a balance between what is taken in and given out over a long period of time (Fig. 11:11).

The recycling of oxygen and carbon

The two major processes involved in recycling these substances are photosynthesis and respiration. In photosynthesis oxygen is evolved as a by-product and this is the only source of oxygen which can be used for the respiratory break-down of food by both plants and animals. Over millions of years a balance has been achieved, between the amount of oxygen released by green plants and the amount used up in respiration and combustion. The same happens on a small scale in a 'bottled garden'.

The carbon dioxide in the atmosphere is the source of carbon for all plants and animals. The only method of extracting it is through photosynthesis, when it is built up into carbohydrates and later into other organic compounds such as fats and proteins. These substances are passed from the producers to the consumers through the food webs. At each stage they are digested and the products are built up into the organic substances of the organism's own body. But how is carbon dioxide passed back into the atmosphere to keep up the supply? There are three main processes which bring this about:

1. During the respiration of plants and animals some organic substances are broken down for energy release and carbon dioxide is given out.
2. During decay, when these organic substances are broken down by bacteria and fungi, there is a release of carbon dioxide.
3. During the combustion of carbon compounds such as wood, peat, coal and petroleum. We can think of these substances as reservoirs of carbon stored up over varying lengths of time, in the case of coal and petroleum, for many millions of years.

One of the problems facing man today is whether this cycle is being upset by the enormous amount of fuel combustion that is taking place to provide energy for industry and transport—particularly the motor car. These processes are putting back immense quantities of carbon dioxide into the atmosphere, the long term consequences of which are uncertain (p. 157).

The recycling of carbon is summarised in Fig. 11:12).

The recycling of nitrogen

The nitrates in the soil are the key substances in this cycle because from them plants build up the proteins which are essential for life and from which all animals derive their proteins. The supply of nitrates in the soil would quickly be exhausted if it were not for several important processes going on continuously.

1. During decay the proteins of dead plants or animals are broken down in stages. First they are turned into ammonium compounds by certain kinds of **saprophytic** bacteria. Then the ammonium compounds are acted upon by **nitrifying** bacteria of two kinds, one kind oxidising them into nitrites and the other oxidising the nitrites further into nitrates.
2. During excretion nitrogenous waste in the form of various ammonium compounds is returned to the soil or water where it is then acted on by the nitrifying bacteria as above.
3. Certain bacteria can extract atmospheric nitrogen and build up nitrates from it. They are called **nitrogen-fixing** bacteria. Some species live freely in the soil while others live in nodules which grow as swellings on the roots of plants belonging to the *Papilionaceae*—a family which includes clover, beans, peas, vetches and sanfoin (alfalfa) (Fig. 11:14). Each nodule contains millions of these bacteria living in **symbiosis** with the host plant. Symbiosis is a relationship between two organisms when they live together in harmony, each benefiting the other. In this case the host plant gains nitrogen compounds which the bacteria synthesise and the bacteria receive sugars made by the host plant.

Farmers make use of these plants with their nitrogen-fixing bacteria by growing them first as a hay crop and then ploughing in the roots; this increases the nitrates in the soil.

Unfortunately for the farmer some soils, particularly those that have become sour (acid), contain **denitrifying** bacteria which have the

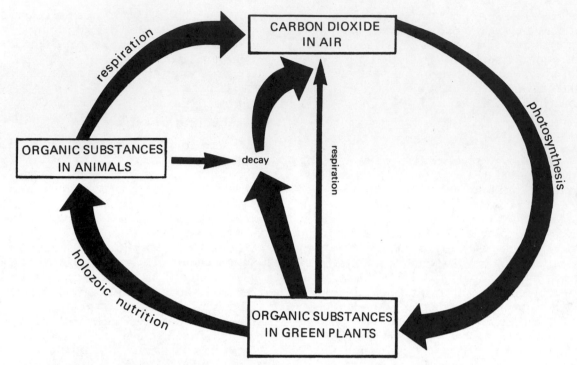

Fig. 11:12 Diagram showing the recycling of carbon.

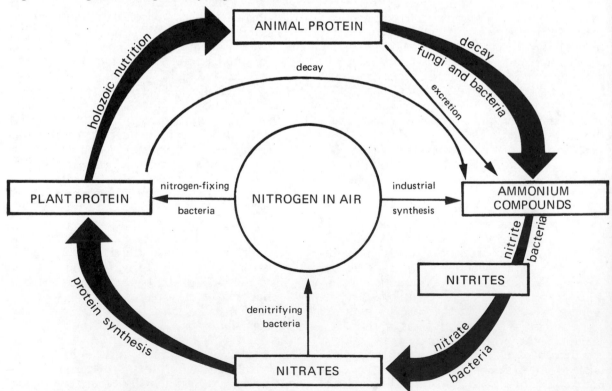

Fig. 11:13 Diagram showing the recycling of nitrogen.

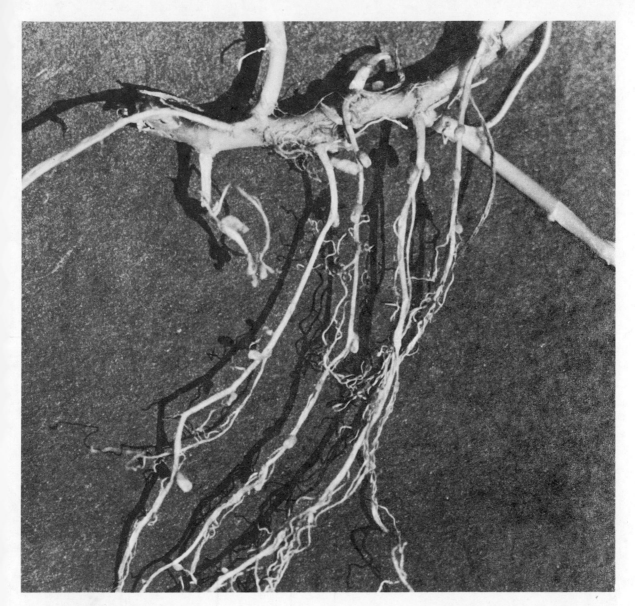

Fig. 11:14 Root nodules on clover.

opposite effect, acting on the nitrates and releasing the nitrogen from them into the atmosphere.

Although air contains about 79% nitrogen, only the nitrogen-fixing bacteria can use this potentially large source. However, by an extremely important industrial process this nitrogen is combined with hydrogen to form ammonia from which such artificial fertilisers as sulphate of ammonia are formed. These fertilisers are used to replace the nitrates lost to the soil when crops are grown. In a natural ecosystem, nitrates used by plants are replaced when the plant dies or decays, but with food crops the product is removed and so nothing is returned. Because of this, manure or fertilisers have to be added.

The details of recycling of nitrogen are summarised in Fig. 11:13.

12

The soil as an ecosystem

How to study an ecosystem

So far we have studied some of the theoretical aspects of ecosystems; we can now use these principles as a basis for studying an ecosystem for ourselves. To do so we might follow four lines of investigation which could be expressed by these questions:

1. What is the physical environment like as a place for organisms to live in?
2. What kinds of organisms live there? How are they distributed and in what numbers?
3. How are these organisms adapted in structure and behaviour for living there?
4. What interactions are there between the organisms in terms of food relationships, competition, etc.?

Let us study soil as an example of an ecosystem.

THE SOIL

How can soil be analysed?

What is soil like as a place to live in? For organisms to survive, the soil must contain adequate food, water and air and it must be in a form which allows it to be penetrated by animals and by the roots of plants. Soils differ greatly in composition and these differences determine the organisms which can live there. Therefore we will take two contrasting types—sandy and clay soils—and analyse them to determine to what extent they provide the necessities for life. Working in pairs, one of you should analyse the sandy soil, and the other the clay.

Each sample should be obtained from the top 15cm and placed in a polythene bag to prevent evaporation of water.

a) What solid matter is it made up of?

Quarter fill with soil a tall narrow jar with a screw top, or a gas jar with a tightly fitting cork; add water until the jar is about three-quarters full, put on the top, shake vigorously and leave on the bench to settle. If the soil contains particles of various sizes, the largest and heaviest will settle first and you should see a gradation (Fig. 12:1). Soil particles are classified according to size: those above 2mm in diameter, gravel; 0·1–2mm, sand; less than 0·1mm silt and clay. The latter is so fine that it may take a long time to settle. Obtain a rough measurement of the depth of the various layers corresponding to the three categories given above. How different are the two samples? Are there any particles which float? These will probably comprise the dead remains of plants, collectively called humus.

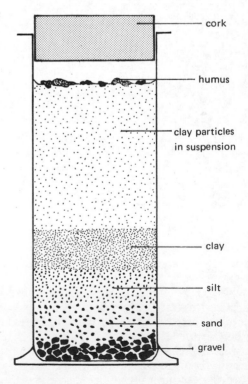

Fig. 12:1 A quick method of determining the composition of soil.

138

b) How much air is present in the two samples?

To obtain an accurate figure, the analysis should be done on undisturbed soil, but a rough estimate, using your sample, may be made as follows:

> Place about 50cm³ of soil in a measuring cylinder, tap it gently on the bench so that the soil consolidates and top up with more to reach the 50cm³ mark again. Now add 50cm³ of water and stir gently with a stiff piece of wire until all the air bubbles have been released from between the soil particles. Measure the final level of the water. The difference between this level and 100cm³ is the amount of air in 50cm³ of soil. Which sample would provide most oxygen for soil animals and plant roots?

c) Are there soluble substances dissolved in the soil water?

You could find out in this way:

> Add some distilled water to the soil sample, shake, filter and evaporate the filtrate to dryness in an evaporating dish. Is there a residue left behind? If so, this could contain the mineral salts which may be taken up by plant roots through the root hairs. Sodium, for example, could be tested by putting some of the residue on to the end of a platinum or nichrome wire and placing it in the flame of a bunsen burner. A yellow flame would indicate that sodium was present.

d) How much water is there in the two samples?

The water content of any soil sample varies according to weather conditions; just after rain, it obviously contains much more than during a drought. If a soil sample is spread out in a warm, dry place for several days, all the water in the air spaces between the particles will evaporate and the weight of the soil will become constant. There is, however, some further water in the soil which can only be removed by heating the sample at a temperature of about 100°C. To simplify our analysis we will measure the total amount of water in the two soil samples by evaporating *all* the water in an oven instead of doing so in two steps.

> Place about 15g of soil in a weighed crucible and weigh again. Place the crucible in an oven at not more than 100°C for 24 hours to evaporate the water, cool in a desiccator and weigh again. Put it back in the oven for another hour, cool, re-weigh and, if necessary, repeat the heating until the weight is constant. The difference between the first and last weighings will be the total weight of water that was in the soil samples. Keep the oven-dried soil for the next experiment.

e) How much humus is there in the soil?

Many soil animals, e.g. earthworms, feed on humus, and as it decays, nutrients from it are returned to the soil to be used once more by plants, so the amount of humus present is one of the factors determining which organisms, and how many, can live there.

> Place the crucible containing the oven-dried soil on a wire gauze and heat it strongly with a bunsen. The humus will burn up completely into carbon dioxide and water vapour. After 15 minutes, cool in a desiccator and weigh. Heat it again, cool and weigh once more; repeat if necessary until the weight is constant. Any loss in weight represents the weight of humus. Which of the samples contains the most humus? The matter that remains is the mineral content of the soil—the actual rock particles. Compare the weight and size of particles in the two samples.

f) How quickly does water drain through soil?

If drainage is fast, this is important for two reasons: 1. Soluble salts in the soil, including artificial fertilisers, will quickly be removed from the soil—for example, after heavy rain. The process is known as **leaching**. 2. There will have to be a greater supply of water, e.g. a higher rainfall, to provide the organisms with all they need as the top soil will dry out quickly. You can compare the drainage rates for the clay and sandy soil in this way:

> Fit up two funnels of equal size (Fig. 12:2) and plug each with glass wool. Add some clay soil to one funnel and an equal volume of sandy soil to the other. Pour on enough water to soak both lots of soil and keep adding it until it is

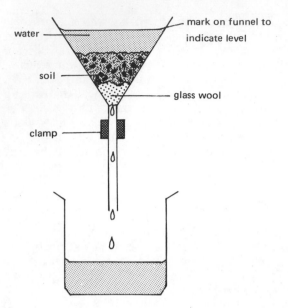

Fig. 12:2 Apparatus for comparing the drainage rates of water through different soils.

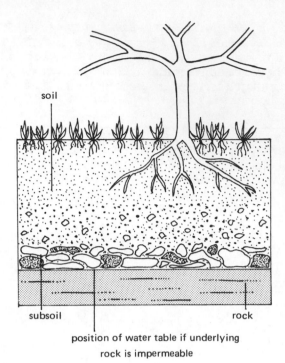

Fig. 12:3 Profile of soil and underlying rock.

dripping regularly from the base. To find out how much water soaks through in a given time, place a beaker under each simultaneously and keep topping up the funnels so that the levels of water above the soil remain constant (why?). When enough water has collected, remove both beakers at the same time and measure the amounts of water in each in a measuring cylinder. The difference between the two indicates the relative drainage rates of the two soil samples.

g) Soil capillarity

Water can be drawn towards the surface of the soil as well as drain down through it. After rain some of the water is retained between the soil particles, but the excess drains away through any rocks which are porous until it finally comes to an impervious layer of rock. Unable to sink any further, it accumulates to form the underground water table (Fig. 12:3). Plants can tap this reservoir even if their roots cannot reach the water table itself, because it is drawn upwards by capillarity through the spaces between the soil particles. Which of the two soil samples will be most efficient in this respect? You can find out in the following way:

Plug two glass tubes about 1·5cm in diameter with glass wool and then fill them with oven-

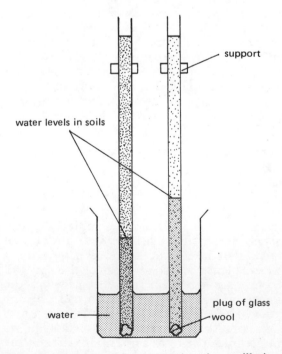

Fig. 12:4 Apparatus for comparing the capillarity of different soils.

dried powdered clay and sand samples respectively (Fig. 12:4). Ram the soil down as hard as possible. Now stand them in a beaker of water. You should be able to tell how high the water travels by the change in colour of the soil. Note the level reached in each tube every 10 minutes for the first hour and again the next day. In which does the water rise fastest to begin with? In which tube does the water reach the greatest height eventually?

We are now in a position to summarise the main differences between a sandy and a clay soil.

	Sand	Clay
Size of particles	Large	Small
Air content	Large air spaces	Small air spaces
Drainage	Good	Poor, i.e. easily waterlogged
Retention of water	Poor	Good
Nutrients	Easily leached out	Retained longer

If you were a farmer you would also add that a sandy soil is easy to plough and warms up quickly in the spring, so producing earlier crops; in contrast, a clay soil is heavy and difficult to work and takes longer to warm up.

If you consider these properties from the point of view of the animals and plants living in these soils, with their need for oxygen, food and water, you will see that both soils have good and bad characteristics. It is therefore not surprising that some of the best soils consist of a mixture of both clay and sand, together with a good humus content; these soils are called **loams**.

Soil texture

Apart from its component parts, the suitability of a loam as a place in which organisms can live depends upon its texture. This results from the manner in which it is built up. A good soil texture is one which is crumbly; when pressed together between the fingers it does not go into a paste but separates into small **crumbs**. These are aggregates of rock particles of various sizes incorporated with humus; they are held together by the surface tension of the water film which lines the air spaces. The larger air spaces between these crumbs of soil allow rootlets to penetrate and animals to move.

Another factor which affects the organisms which live in soil is the chemical nature of its constituents. We will take one example:

The presence or absence of lime in the soil

A soil containing a significant amount of calcium carbonate or lime is said to be **calcareous**, one lacking in lime, **non-calcareous**. Each kind has its characteristic flora, because some plants, e.g. wild clematis, dogwood and spindle only grow well when it is present and others, such as heather, cannot tolerate much lime.

Calcareous soils are usually derived from underlying rocks, such as chalk and limestone by a process of weathering. Soils formed in this way are said to be **sedentary**. However, **sedimentary** or **alluvial** soils (formed from particles brought down by rivers or glaciers) may also contain calcareous particles if the river or glacier passed over calcareous rocks during its course. You can find out if your soil sample is calcareous in the following way:

Add dilute hydrochloric acid to the soil in a test tube. If calcium carbonate is present, carbon dioxide will be given off as bubbles. How would you test that it was carbon dioxide?

Calcareous soils are neutral or slightly alkaline (you can test this using universal indicator), but non-calcareous soils tend to be acid. Acid soils usually develop because of insufficient oxygen. In moorland, for example, the soil is poorly aerated because of its waterlogged condition. This means that aerobic bacteria are unable to flourish, so the humus is not broken down completely and collects in the form of peat. Partial breakdown of humus results in the formation of humic acids. This accounts both for the acid nature of the soil and for its lack of fertility, as potential nutrients in the humus are not returned to the soil. Heavy clay soils also tend to be acid for lack of adequate aeration.

Acid soils can be greatly improved by adding lime to the soil (usually as slaked lime). This has the effect of neutralising the humic acids. But lime also improves the texture of the soil.

When worked into a clay soil it causes the clay particles to clump together into larger units—a process called **flocculation**. This makes the soil lighter and more easily worked and allows air to penetrate better. This in turn improves conditions for the aerobic bacteria which act on the humus. The process of flocculation may be demonstrated as follows:

Take two beakers of equal size and add $10cm^3$ of powdered clay and $150cm^3$ of water to each. Add $2cm^3$ of lime to one beaker only. Stir both beakers vigorously with a glass rod and then allow the contents to settle. The speed of settling will vary according to the size of the particles. In which beaker does the clay settle first?

Soil factors

We can summarise our findings so far by saying that all soil is made up of five non-living constituents—mineral particles, water, soluble mineral salts, humus and air—but it varies considerably according to the chemical nature of its constituents, the size of the particles and the relative proportions of its component parts. Soil organisms have to be adapted to all these factors, but there are also other physical factors which are characteristic of soils to which organisms also need to be adapted. These are: 1. A relatively low temperature (even if the surface of the soil becomes hot in strong sun, the warmth does not penetrate far). 2. A high humidity. 3. Absence of light except at the surface. We must take these factors into account when we study the organisms themselves.

Plants living in the soil

Owing to the absence of light below the soil surface no green plants can exist there, so the flora consists of teeming masses of bacteria and fungi which feed saprophytically on the humus. You have seen some of these for yourself (Book 1 p. 191). However, most soils if left undisturbed for long enough become covered in green plants whose roots penetrate into it. In places such as woods and grassland, for example, this layer of vegetation is so thick and composed of so many species that it is difficult to understand how they can all find enough space and nutrients.

Examine the root systems of four of the commoner plants found living together in any one well-established habitat, such as a wood or grassland. Dig them up carefully with plenty of surrounding soil and soak them in water before washing off the soil that still clings to the roots. Compare the lengths of the roots and the general lay-out of the root systems: this should help you to explain why certain species can live close together and yet obtain enough nutrients. There are, of course, other reasons which enable them to do this which do not concern their roots; can you think of any?

Finding soil animals

Soil animals vary greatly in size, but the majority are very small. Could small size be an adaptation to living in soil? Soil animals also vary greatly in their distribution; some are mainly surface-dwellers, others occur at various levels within the soil. Let us consider the surface-dwellers first.

Surface-dwellers

If a soil organism is adapted for living in a dark, damp habitat where temperatures are low, life on the surface of the soil could be difficult, as this is a region which is exposed to light, tends to be drier and is subject to greater changes of temperature; it is also exposed to the larger predators, e.g. birds. But the soil surface is seldom uniform; in some places it is covered with leaves, in others with large stones or logs and in most places there is vegetation to provide shelter. You will remember from your choice-chamber experiments on woodlice that this species tended to move into dark and humid places (p. 116). It is therefore probable that most soil animals will be found by searching in any place where these conditions are prevalent. Also, one would expect these animals to be more active at night.

Keep these points in mind when searching for the animals, and collect representative specimens of each species for identification and examination. Keep each in a specimen tube and put in a few damp leaves to give them moisture and something to cling to. Add a label indicating where each specimen was found. Here are some methods that you could use for finding them:

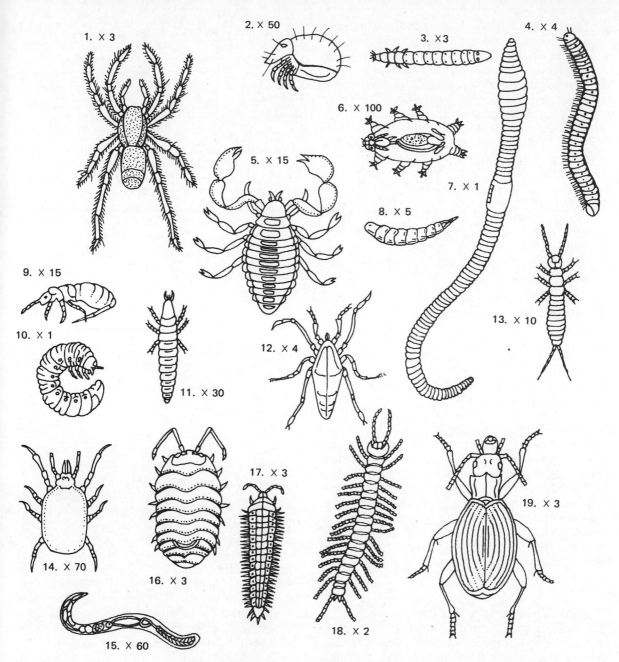

Fig. 12:5 A selection of soil organisms, mainly surface-dwellers: 1 Wolf spider. 2 Armadillo mite. 3 Wire worm (click beetle larva). 4 Snake millipede. 5 False scorpion. 6 Tardigrade. 7 Earthworm. 8 Fly maggot. 9 Springtail. 10 Larva of cockchafer beetle. 11 Proturan. 12 Harvestman. 13 Dipluran. 14 Mite. 15 Nematode. 16 Pill woodlouse. 17 Flat millipede. 18 Centipede. 19 Ground beetle.

1. Search the soil surface in various habitats in the daytime; look under any movable object on the surface including the leaves of the larger plants.
2. Spread out some wet sacking on the ground overnight and look for animals underneath it the following morning.
3. Sink jam jars up to their rims in soil to act as pit-fall traps. Examine them morning and evening. This should enable you to find out when surface-dwellers are most active.
4. Collect leaf litter from a wood or under a hedge. Use a hand fork to scrape together a large heap; put it into a polythene bag and then sieve it, a few handfuls at a time, through a wide-meshed sieve on to a white sheet.

Do your findings help to confirm that surface-dwellers tend to be more active at night, while in the day they seek moist, dark, cool and sheltered places?

Identify your specimens according to their main types; the illustration in Fig. 12:5 may help you, but also consult appropriate books. (See Appendix). Make a table, stating where each was found, together with any structural adaptations that you observe which would help them to live as surface-dwellers. Consider such aspects as shape, size and organs of locomotion. Much time would be needed to discover their feeding relationships—whether they were vegetarians, scavengers or predators—but you could devise simple feeding experiments to test their reactions to various foods. Each member of the class could study one species, and then the results should be pooled.

Animals living in the soil itself

The larger species can be found by digging, and animals such as earthworms can be drawn to the surface by soaking the soil with 2% formalin (Book 1 p. 38), but the majority of soil animals are too small to find in these ways. One technique that can be used relies on the principle that if soil organisms are adapted to moist, dark, cool conditions, they can be driven out of soil by illuminating it and making it dry and warm. However, the treatment must not be too severe, otherwise they will die before they come out of the soil.

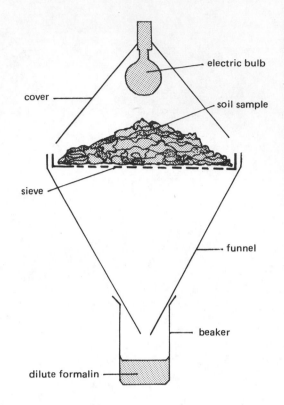

Fig. 12:6 Tullgren funnel.

The **Tullgren apparatus** consists of a large funnel (Fig. 12:6) made of metal, glass or tin foil, and a sieve (on which the soil sample is placed) with a mesh fine enough to prevent the soil from slipping through, but large enough for the animals to penetrate. An electric light bulb fitted above the funnel supplies both a source of light and heat, and so will dry out the sample as well. The animals, in moving away from these adverse conditions should pass downwards out of the soil and fall into the beaker of dilute formalin below, which will kill and preserve them.

The class should set up at least three of these pieces of apparatus using soil samples from different places—the top soils of woodland or grassland contain a large variety of organisms. Break up the soil gently before placing it on the sieve so that the animals may escape more easily. Switch on the lamp and leave for several hours.

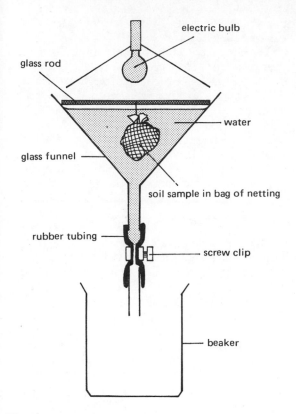

Fig. 12:7 Baermann funnel.

you to compare the numbers and the variety of organisms *either* from different top soils *or* from the top 6cm of one soil compared with the next 6cm. To do this, similar pieces of apparatus should be used and equal quantities of soil taken. The soil can be extracted with a bulb corer (Fig. 12:8); this will enable you to remove equal volumes of soil from the regions to be compared.

The Baermann apparatus

This method is used to extract soil animals which live in the soil water and which do not react well to the Tullgren method. The principle is to supply top lighting and heating as before, so that the animals leave the soil, but to prevent the soil from drying up, the soil sample is surrounded with water.

The class should set up a corresponding number of these pieces of apparatus to those in the last experiment and soil samples from the same regions should be tested. This will enable you to compare the efficiency of these two techniques for extracting different species of the soil fauna.

Enclose each sample in a piece of curtain netting and suspend it in the water as in Fig. 12:7. Switch on the lamp and leave it for three hours. By this time most of the organisms will have left the soil and will be seen mainly in the neck of the funnel. Place a beaker underneath, open the clip and run off enough liquid to remove the animals. Examine them under the microscope and identify the main kinds as before.

You will probably find large numbers of **nematodes**, tiny worm-like creatures pointed at both ends. They are extracted very efficiently by this method. How would you adapt this extraction technique to estimate the *number* of nematodes present in the top soil of a square metre of grassland?

Examine your catch under a hand lens or microscope and try to identify the specimens with the help of Fig. 12:5 and appropriate keys and books. Make a table listing the main types, and for each animal, note any structural features which you think might help it to live in the soil, as you did for the surface-dwellers. Are these adaptations similar to, or different from, those you found for surface-dwellers? Can you think of an explanation for any differences?

By using Tullgren funnels, it is possible for

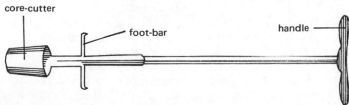

Fig. 12:8 Bulb-corer.

These two methods of extracting organisms from soil should have given you some idea both of the large number of organisms which exist in soil and of their diversity. From the various methods you have used to find and extract soil organisms, and from your observations on their structural and behavioural adaptations, you should now have some idea of the complexity of this ecosystem and the vast number of organisms comprising it. You will not have been able to work out many of the details of the food relationships of the various species, but the following points should help you to fill in some of the gaps and see how the soil organisms are all inter-related.

1. The only producers are the green plants.
2. The available plant food comes from living roots or the dead remains of plants (humus).
3. The scavengers include earthworms, millipedes, woodlice and the minute springtails.
4. The main decomposers—bacteria and fungi —and the nitrogen-fixing and nitrifying bacteria occur in soil in vast numbers.
5. The smaller carnivores include ground beetles, centipedes, spiders and certain mites.
6. The larger carnivores include moles below the surface and shrews and blackbirds above. These, with many others, link the soil with other ecosystems.
7. When animals other than those in the soil die, their bodies are eaten or decomposed and may add food material for soil organisms. The same also applies to their waste products—dung and urine.

> From these facts and other data the class has collected, try to construct a diagram to show some of the inter-relationships of soil organisms.

CHANGES WITHIN AN ECOSYSTEM

If it were possible to count the numbers of the various organisms present in an ecosystem such as soil over a period of years, we should probably find that although numbers varied from season to season and from year to year, they would tend to fluctuate around a mean, i.e. average numbers would be maintained over a long period. This is characteristic of a **stable** ecosystem.

Temperate woodland and tropical forest are other examples of stable ecosystems, because if left undisturbed by man conditions remain fairly constant and their communities remain in balance.

But drastic things can happen to what appears to be a stable ecosystem. Volcanic eruptions, floods, droughts, hurricanes, etc., can greatly alter an ecosystem or destroy it completely. What happens then?

Colonisation and succession

An ecosystem which has been drastically changed undergoes **colonisation**. Take, for example, an area of forest destroyed by fire. The first plants to re-appear will probably have originated from spores or seeds brought by various means, especially wind; mosses and grasses may be among them. As a result, the ground will soon be covered by herbaceous plants which compete with each other for light and space. Later, the slower-growing shrubs and trees will make their appearance and as they grow taller will shade the ground flora, much of which will die out for lack of adequate light. Later still, the trees will outstrip the shrubs and in turn shade them; in consequence some of the shrubs will die if the trees are close together, and so the original forest structure is restored. In this way recolonisation of a devastated area tends to produce in the end what is called a **climax** vegetation; this is a relatively stable ecosystem.

This series of changes is known as **succession** and, of course, it applies equally to the fauna which is dependent upon the plant life. Thus during this process of succession every organism which succeeds in establishing itself changes the conditions and makes the habitat either more or less suitable for other organisms; their success or failure in turn will change the conditions again until eventually, in the climax condition, a balance will be established which is relatively stable.

If man interferes with a habitat by tree-felling, ploughing, etc., and the region is then left alone, it gradually changes through a succession of communities until the original climax condition is produced. So if the human population of southern England was com-

Fig. 12:9 Pond ecosystem showing gradual colonisation of open water by marsh plants.

pletely evacuated, most of the region over the following 50 years would revert to deciduous temperate forest. The same would apply to all the major world ecosystems such as tundra, savannah and tropical forest. You will see later (p. 152) how man has to battle constantly with nature to prevent this from happening, if it is in his interests to do so.

You can observe succession taking place in this way:

Dig a square metre of soil very thoroughly, carefully removing all roots and whole plants. Leave it completely alone for several months. At regular intervals count the numbers of individual plants of each species present and note the changes that occur. To reach a climax you would have to keep this up for 50 years or more, but many changes occur, even over a few months, as new arrivals compete for light and space and vegetarians such as slugs and snails take their toll.

Changes in population

The term **population density** is used in the same sense for animal and plant numbers as for humans. It refers to the number of a particular species in a given area.

In any ecosystem the population of a particular species depends upon four factors: 1. Natality—those being added as a result of reproduction. 2. Mortality—those being removed from it by dying. 3. Immigration—those reaching it from other ecosystems. 4. Emigration—those leaving it for other ecosystems. The population density will be stable if:

natality + immigration = mortality + emigration

In practice, all populations fluctuate because of the ever-changing conditions within an ecosystem which affect these four components

147

of the equation. Sometimes the fluctuations are cyclic, i.e. the rises and falls are fairly regular and the average population density over a long period of time is maintained. Thus the snow-shoe hare population in Canada shows high or low peaks about every ten years (Fig. 12:10) and the Scandinavian lemming every three or four. However, most fluctuations are much more irregular.

In all species there is the potentiality to increase in numbers. A pair of rabbits under perfect conditions could theoretically produce a population of 14 million in 5 years, but in practice this does not happen because of the effects of a number of factors. Let us consider some of these:

a) Climatic factors
Extremes of weather in terms of temperature, rainfall, wind, humidity, etc., combine in various ways to affect the density of the population. Some of these are cyclic, such as the seasonal effects of summer and winter in temperate regions and the wet and dry seasons in tropical countries. But when these factors are extreme a high proportion of a population may be destroyed and it is many years before

more normal numbers are restored. For example, in England, a period of unusually prolonged frost in 1962 caused the ground to become so hard that bird species which were largely dependent upon soil fauna for their food, e.g. green woodpeckers, were greatly reduced in numbers. Similarly, periods of severe drought have caused the deaths of vast numbers of ungulates on the African savannah. You will think of many other examples of the effect of climatic conditions on population densities.

b) Food supply
When food is abundant relative to the number of individuals in a population, it is of no importance in regulating population density, but when scarce it will cause numbers to drop, either because competition for what remains will increase and this may lead to starvation or death from other causes, or because there will be emigration to areas of more plentiful food supply. Insectivorous birds such as swallows, swifts and warblers migrate from temperate regions as winter approaches to tropical countries where insects are more abundant, and return in the spring when insects are available once more in sufficient numbers.

c) Space
When the population density is low the individuals are spread out and there is enough space for each to acquire sufficient food and shelter to live successfully, but under crowded conditions, competition for food and shelter becomes greater and, in many species, behaviour becomes more aggressive. This happens in small rodents such as voles which, like snow-shoe hares, have periodic increases in population followed by dramatic crashes in numbers. In this example one of the important factors in causing this sudden drop, is stress (p. 110). When overcrowding occurs the animals are constantly being stimulated by their neighbours. There is more fighting, mothers with litters cannot care for their young adequately and there is insufficient time for resting and feeding. This constant unrest upsets the normal working of the endocrine system, the pituitary and adrenal glands become gradually exhausted and the animals die of 'shock'. After the crash in numbers, the few survivors have plenty of space, so the population builds up again and the cycle is repeated.

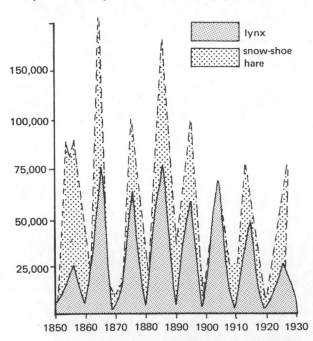

Fig. 12:10 Graph of snow-shoe hare and lynx population over 80 years based on skins obtained by the Hudson's Bay Company of Canada.

148

You might consider the effects of overcrowding on man. Does it cause more aggressive behaviour and increasing stress symptoms? Compare life in overcrowded cities with life in the country; in overcrowded schools with those possessing much greater space.

Many species avoid overcrowding by showing territorial behaviour. A **territory** (Book 1 p. 106) is a part of the habitat over which an individual, pair or social group asserts a dominating influence over other members of the same species. It varies greatly in size according to the food and breeding requirements of the species. In birds, territories are defended against intruders by various forms of aggressive behaviour such as singing and displaying; in mammals, ritual fighting may occur.

Territories provide some kind of home for breeding and a refuge from enemies. The number of adequate shelters in an area often influences the density of the population. For example, it has been found that by increasing the number of nesting sites in a wood by putting up nesting boxes, the population of tits and flycatchers has increased. Conversely, when old trees have been cut down, the number of natural nesting holes has been reduced and numbers have fallen.

So the number of potential homes, as well as the available food, will influence the size of the territory and thus the population density.

d) Diseases and parasites

Epidemics may sweep through both animal and human populations and cause numbers to fall. For example, when myxomatosis was introduced into Britain over 90% of the rabbit population was destroyed. But even when some degree of immunity has been established, disease is an important factor in limiting numbers. Bovine tuberculosis, for example, is common in buffalo herds in East Africa and even if it does not cause death it will weaken the animals and so make them easier prey for lions. This is an example of how different factors often interact to affect numbers.

Parasites also play an important role in regulating numbers in some species. We have already seen an example of this with the large white butterfly and its ichneumon parasite (Book 1 p. 68). When a population reaches a high level, it is much easier for the parasite to spread from one host to another, and the same applies to the organisms causing infectious diseases. This leads to the important principle that factors such as available food, space, shelter, parasites and disease exert an increasing influence as population density becomes greater. Is this also true for the human population?

e) Man

Man is one of the most important factors influencing animal and plant population densities because he has drastically manipulated the environment for his own ends. The result is that many of the world's ecosystems are no longer in a relatively stable condition and many artificial ones have been formed. We shall consider the implications of this in the next chapter.

13

Man and his environment

Man's brain has developed to an incredible extent over the past 100,000 years. This has allowed him to become more independent of animal instincts and more capable of exploratory and creative actions, but at the same time it has increased his powers of destruction. Through the development of these powers man has increasingly destroyed or modified much of the natural environment and created an artificial one of his own making.

Past history

In order to understand man's special relationship with his environment it is necessary to review his more recent evolution briefly. Some 15,000 years ago man was an unobtrusive species within the living world; his effect on the environment was relatively small and he was well adapted, like other members of the fauna, to the ecosystem of which he was part. But his brain was developing fast and he was becoming much more effective in making stone implements, so he began to make a greater impact on his environment. He started to cut down the forests and grow crops, build more durable huts and domesticate animals. This meant that he gradually changed from a nomad seeking food wherever it could be found to a village dweller surrounded by the fields he could till, the crops he could harvest and where his domesticated animals could be cared for and protected. He learnt how to store the grain that he grew and preserve the fish and meat he procured to tide him over

any periods of scarcity. He improved his method of communication with other individuals by inventing language; this made co-operative action much more effective and started him along the road of **cultural inheritance**—the passing on of experience from one generation to the next. He discovered the value of fire for cooking and preserving, for clearing land before planting, for keeping warm in colder regions and, later on, for moulding metals to form more useful tools for hunting, for war and for agriculture. So man moved from the Stone Age through the Bronze Age to the Iron Age. Life became increasingly complex—there were more things to be done, more trees to be felled, huts to be built, animals to be cared for, fields to be tilled and tools to be made; also, there was pottery to be moulded and baskets to be woven. Thus man became more specialised; there was greater division of labour, bartering of products took place, roads were built to link up the villages and as man began to prosper, larger towns and cities were constructed. Our modern world is the logical development of these trends.

Population growth

With greater powers for subduing nature and harnessing natural resources for his own ends, man has increased in numbers. At first the growth rate was slow. It is impossible to estimate numbers in the distant past with any accuracy, but it is thought that the 100 million mark was not reached until about 3000 B.C. The population probably reached 500 million by A.D. 1650 and 1000 million around the middle of the 19th century. But since then the rate has accelerated to an alarming extent, due largely to man's success in tackling diseases such as malaria, cholera, plague and typhoid (Book 1 Ch. 7) which in the past had been major killers. So by 1930 numbers had risen to about 2000 million, by 1960 to 3000 and by 1970 to 3500 million. It is estimated that by the end of the century, when you are in middle life, it may be in the region of 6500 million! (Fig. 13:1). This means that at the present time babies are being born all over the world at a rate approaching 200 a minute while people are dying at the much slower rate of about 110 a minute. Of course, this rate of

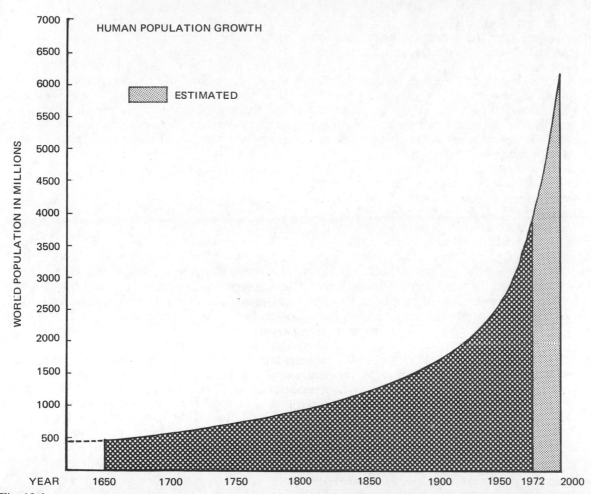

HUMAN POPULATION GROWTH

ESTIMATED

WORLD POPULATION IN MILLIONS

7000
6500
6000
5500
5000
4500
4000
3500
3000
2500
2000
1500
1000
500

YEAR   1650   1700   1750   1800   1850   1900   1950 1972 2000

Fig. 13:1

Fig. 13:2 Overcrowding—a scene on Margate beach.

151

increase varies greatly from one country to another. In Brazil, for example, the population will have doubled in about 20 years. Think what this means in terms of food, water, housing, power and other amenities. All these will have to double in 20 years just to keep pace with the present standard of living. Even in countries where the increase is low, as in parts of Western Europe, the problem is great as these countries are already very overcrowded.

Another factor of great importance is that, all over the world, vast numbers of people have moved from the land to urban communities, in the hope of obtaining a better living. This has led to great overcrowding and poverty in many great cities, with hopelessly inadequate housing, food, water and sanitation, and all the social problems that go with them. Today, about one-third of the world's population lives in towns of over 20,000 people and by the end of the century it will be more than half. In highly developed countries the proportion is already much greater. This staggering increase in numbers is already having an immense effect on people everywhere. When and how will the increase level off? Clearly some means of regulating numbers is essential (p. 35).

The effects of the population explosion, together with man's technical achievements,

Fig. 13:3 Large-scale monoculture of cotton.

have already led to the exploitation of the natural environment on an unprecedented scale. But fortunately there is now a greater awareness of the environmental problems facing mankind and much is already known about how these may be tackled. What is needed most is the determination for all people and nations to work together so that our knowledge is applied sensibly. Let us now examine some of the effects of man on his environment from an ecological point of view.

THE EFFECT OF AGRICULTURE

When a forest is cut down and a food crop is grown in its place, a natural climax ecosystem with its vast number of species in a state of dynamic equilibrium is replaced by a **monoculture**, i.e. an unnatural concentration of a single species in one area. Look around a farm and you will see crops of various kinds grown in different fields—some to provide cereals or roots, others grass for domestic animals. Whatever the crop, it is an unstable community and if left on its own would revert to the natural climax ecosystem once more.

You know what happens when a garden is left untouched for a long time; it soon becomes a wilderness. Only by constant attention can it be kept as it is wanted. The same applies to any land which is farmed; pasture is only kept as such by routine grazing or cutting for hay; fields can only be maintained in a fit state for growing crops by regular cultural methods. The battle to prevent the natural succession from taking place brings with it many problems.

Pests and diseases

By growing crops in large concentrations man provides ideal conditions for the pests which feed on them, and the parasites, particularly fungi, which cause disease. Here is food in abundance and excellent conditions for diseases to spread from one plant to another. Consequently, multiplication is rapid and the resulting damage very great. To avoid this happening man has tried to eliminate these competitors for his crops by using toxic chemicals. Many of them have been very effective, but their use has also created new problems.

Pesticides

The perfect pesticide is one which destroys a particular pest and is completely harmless to every other form of life; no such pesticide exists or is likely to. Do you see why?

In judging the usefulness of a pesticide it is necessary to take into account both its effectiveness and the possible harm it does to other forms of life, including ourselves.

Beneficial effects

The value of such substances as DDT has been very great in controlling disease and increasing food output. Their use has greatly reduced the incidence of such killers as malaria, yellow fever, plague and typhus by destroying their insect carriers, and when sprayed on crops they have greatly increased the yield. In some countries, due to the elimination of such pests as wireworms, the output of cereal crops has increased by 60% and potatoes by 70%. Also, the spraying of orchards has led to a marked improvement in both the quantity and the quality of fruit.

Harmful effects

Pesticides are often indiscriminate in their action and vast numbers of other animals may be destroyed. Some of these may be the predators which naturally feed on the pests, others may be the food of other animals, thus causing unpredictable changes in food chains and upsetting the balance within the ecosystem. When pesticides are sprayed from the air over large areas of land these effects may be dramatic.

A further danger is that some have a cumulative effect. Pesticides vary in their length of 'life' as toxic substances. Some, such as the organophosphates (and also the herbicides) are said to be **degradable** because they are broken down into harmless substances in a comparatively short time, usually under a year. Others are **non-degradable**, and include those which contain mercury, arsenic or lead. Others, such as the chlorinated hydrocarbons, DDT, aldrin and dieldrin, are extremely difficult to break down and may remain in the soil for over ten years. This has given time for many insects to build up resistance to their action, so research goes on all the time to produce differ-

ent ones. These non-degradable pesticides are potentially dangerous as they accumulate in the bodies of animals and pass right through the food web, being further concentrated at each step until animals at the top of the pyramid may receive enough to do considerable harm. This has been very evident in birds such as herons, which have gradually accumulated in their bodies the DDT they acquired from fish. Similarly, when dieldrin was used as a sheep dip to eliminate parasites such as mites, ticks and blow-flies, the eagles were greatly affected as they fed on any sheep that had died. In both these instances reproduction either ceased, or the eggs had such thin shells that the developing chicks did not survive. The persistence of these chlorinated hydrocarbons may be judged by the fact that so much has been used over the past 25 years that it has found its way from the land and rivers to the oceans and has been dispersed by currents to all regions; even the penguins and polar bears which live literally poles apart have traces in their tissues. Man, being at the end of so many food chains, also accumulates these substances and it is essential to question their possible effect on him. There is no evidence at present that amounts absorbed so far are doing recognisable harm, but the danger of further accumulation must be taken very seriously indeed.

What should be done?

It is easy to say 'Ban all pesticides!' but the pests still have to be kept in check. At the present time pests still claim 10% of the world's food supply even after £600 million has been spent on chemicals to control them every year. If pesticides were totally banned what would happen to the diseases they are controlling and the crops we so vitally need for our growing population?

The problem is not insoluble. Already some of the most harmful substances have been banned in a number of countries and the search continues for alternative methods of pest control. For the present it is obviously a sane policy never to use pesticides unnecessarily, and when they are essential, to use a degradable kind as sparingly as possible. Those of us who have gardens can all play our part in pursuing this policy.

But the long-term answer is to find other effective methods of controlling pests which have far less harmful effects and are based on sound biological principles. This more positive approach is being pursued with vigour. Here are some of the more important methods used:

a) By growing different crops on a particular piece of land in successive years. This rotation of crops reduces the build-up of pests from year to year in that area and so less damage is done.

b) By studying the life histories of the pests, it is possible to sow the crop at a time when least damage will be caused. For example, the larvae of the frit-fly feed on the delicate seedling stages of oats and rye, but do little damage to the older plants; thus by sowing the cereal well before the frit-fly larvae hatch, the damage is much reduced.

c) By introducing a natural predator or parasite of the pest. This is an example of **biological control**. In an ideal situation (which seldom occurs) the organism selected should be one which attacks only the pest species, as otherwise it might turn to another, perhaps useful, species and then become a pest itself! This method has been particularly useful when a plant or animal has been introduced into another country and multiplied excessively to become a pest. In this way the cabbage white butterfly, when it became a pest in New Zealand, was successfully controlled through the breeding and releasing of large numbers of its ichneumon parasite (Book 1 p. 68). In a similar way, buffalo flies are being controlled in northern Australia by dung beetles. These flies breed in cattle dung and multiply very fast. They are a menace because they irritate the cattle so much that meat production falls. Domestic cattle were introduced into Australia but no indigenous dung beetles are present to feed on their dung. In consequence: 1. The dung pats do not decay for months, and as about 200 million pats are produced every day, many acres of pasture are lost each year. 2. The buffalo flies breed in vast numbers on the pats. 3. The nitrogen in the dung is not returned to the soil to enrich it. To deal with this problem dung beetles were imported from Africa and Asia in 1967, and by 1970, 275,000 beetles had been released. One

Fig. 13:4 Female dung beetle rolling dung ball for subsequent burial and egg-laying.

species colonised an area of more than 3000 km^2 within two years. The beetles live on dung which they roll into balls and bury; eggs are laid inside and when the larvae hatch they feed on the dung, pupate and later emerge as beetles; the cycle is then repeated. The beetles can bury a pat within 48 hours, thus preventing buffalo fly eggs from developing. So in one operation they bury the dung, remove the pest and put the nitrogen back into the soil! The beetles can only feed on dung, so there is no likelihood of their becoming pests themselves.

A plant example of biological control is the prickly pear cactus which, when introduced into Australia, thrived so well that it invaded and destroyed vast areas of pasture. By introducing the larvae of a moth which feeds on it, it was brought under control within ten years.

A more controversial method of biological control was used when the myxomatosis virus was introduced to reduce the rabbit population in Australia and later in Britain. Statistically it was very successful, as it caused the deaths of over 90% of the rabbit population which in Australia had reached plague proportions in the absence of natural predators. But on humane grounds the operation was severely criticised as the animals died with distressing symptoms.

d) By rendering the males of a pest species sterile by subjecting them to radiation and then releasing them in the wild population. This has been done successfully with the screw-worm fly which infects cattle. The insects mate normally, but the eggs are infertile.

e) By the development, through breeding experiments, of genetic strains which are resistant to the particular pests. In this way strains of cereals have been developed which are resistant to fungus diseases caused by rusts and mildews. Although this method holds out great promise for many species, it has been found that pests sometimes adapt successfully to the new strains.

Weeds

By ploughing land, man not only provides a good soil for his crops, but he produces conditions which are excellent for weeds. Weeds are often the first colonisers of disturbed soil. They grow fast, form flowers very early, and produce large numbers of seeds; many are annuals, and some even complete their whole life history in a few weeks. Thus they multiply very rapidly and compete strongly with the crop for space and nutrients. There are several ways of lessening this problem:

1. Crops may be sown very thickly. This works well with cereals as it leaves very little space for the weeds to develop.
2. Hand-weeding, hoeing or machine weeding may be done during the early stages of growth. This applies especially to root crops which have to be given more space. The method of weeding varies according to the labour available and the type of crop.
3. Crops may be sprayed with selective weed-killers. This method is particularly effective for cereals, as some of these substances destroy broad-leaved plants (most weeds come into this category) but do not harm grass-like species (cereals).

The removal of crops from the land

In a natural ecosystem plant products are returned to the soil and recycled, but when crops are grown, they are usually harvested and removed, so the soil becomes impoverished. To replace the mineral nutrients removed by the crop the farmer uses artificial fertilisers, but this does not restore the humus content which is so valuable for water retention and soil texture. Previously on mixed farms where animals and crops were produced together, the manure from the animals could be used to enrich the soil for the crops, but with more specialised farming this is no longer possible. Consider the situation where the land is particularly suitable for cereal growing. In recent years new strains have been developed which give very high yields when supplied with large quantities of artificial fertilisers and more and more of these crops are being grown. Results have been spectacular and the countries concerned have benefited greatly. Wheat production in Pakistan, for example, increased from 4 to 8 million tonnes per annum between 1966 and 1971 and in India from 12 to 20 million tonnes during the same period. In south-east Asia rice production has also increased very greatly. In highly developed farming countries great increases have also been achieved, but there can also be dangers from these intensive farming methods and these should be considered.

In East Anglia, for example, in order to grow cereals intensively, hedgerows and wind-breaks have been torn down to increase the size of the fields and to make it easier for larger machines to operate, and the traditional rotation of crops has also been abandoned. Consequently, with the loss of humus, the soil has become lighter and in times of drought some of the top soil has literally been blown away. The use of heavy machinery may also make the soil more compacted and so more liable to flooding. A compacted soil also reduces

Fig. 13:5 Soil blown off the fields into a drainage ditch: East Anglia, 1972.

the amount of fertiliser that reaches the root systems of the crops, and in wet weather the fertiliser may be washed out of the soil and reach rivers and lakes. Here the extra nitrogen and phosphorus cause rapid growth of algae and when these die the bacteria which bring about their decay use up so much oxygen from the water that fish and other aquatic animals may suffocate. This form of pollution is known as **eutrophication**; it is already a very serious problem in lakes and rivers wherever these intensive methods of farming are practised.

Factory farming

The need for more cheaply produced animal food with less land to produce it has led to this specialised form of farming, where animals are reared indoors. The ethics of these methods are not considered here, but the results are certainly impressive. In Britain in 1970, 12 million cattle, 7 million pigs and 127 million poultry spent all or some of their life indoors. With the production of new strains and intensive feeding methods beef cattle now mature in 10–18 months, compared with 3 years in 1946; broiler chickens are ready in under 9 weeks compared with 16–18 weeks; on average a hen lays 211 eggs per year compared with 108, and cows yield 3700 litres of milk compared with 2475. But consider the problems which result:

1. All the animals produce a colossal amount of manure which is concentrated in one place, and in most cases there is no agricultural land nearby to put it on. It is not economical to move it to arable farms which need it so badly and much of it goes into the drains and streams and pollutes them.

2. When large numbers of animals are reared together they are more prone to disease. An epidemic would be disastrous to the farmer, so to prevent it, small quantities of antibiotics are used repeatedly. This action has led to the development of bacterial strains which are resistant to the antibiotics, and this may pose a real threat to the health of both animals and humans.

3. To obtain these impressive yields the animals have to be fed on high protein animal feed, much of which comes from developing countries where the need for extra protein for the human population is far greater. This raises both humanitarian and political issues of great importance.

In order to take a balanced view of all these techniques which man has developed, the beneficial results have to be weighed against the harmful, but in coming to a conclusion it is wise to apply the ecological principle that there should be a true balance between what is taken out of the land and what goes back into it.

This raises big questions—moral, social and political. For example:

1. Should man try to extract from the land the *maximum* possible, irrespective of the harm to the environment the method may cause, or should he be content with less—the *optimum*—which would allow him to keep his land in balance?

2. Do people in the affluent countries actually need all the food that is being produced?

3. How can the developing countries increase their food production without inheriting the same problems as the more highly developed nations?

Fig. 13:6 Factory farming: beef cattle reared indoors.

THE EFFECT OF INDUSTRY AND TECHNOLOGY

The environment has not only been greatly affected by man's agricultural activities, but also by industrial development. The tremendous technological achievements of the past century have brought many benefits, including a rising standard of living, but how far has our environment suffered? How far has industry polluted the air, water and land? How much is being done to keep these essential parts of the ecosystem unharmed?

156

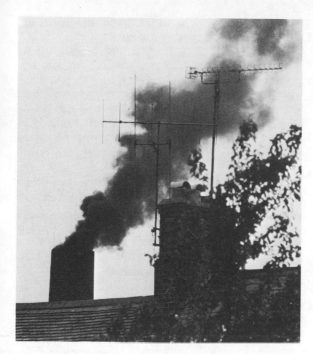

Fig. 13:7 Air pollution: considerably reduced in Britain since the Clean Air Act of 1956.

Air pollution

Industry has been greatly dependent upon combustion to provide energy and wherever fuels are burnt vast quantities of waste pass into the atmosphere. In Britain, before the Clean Air Act of 1956, it was estimated that 2·3 million tonnes of smoke, 5·2 million tonnes of sulphur dioxide, 0·8 million tonnes of grit and 0·3 million tonnes of acid were emitted into the atmosphere each year. These substances had serious effects on plants, causing an annual loss to agriculture of about £10 million, and producing a much more serious effect on the health of animals and in particular, man. Since 1956 the smoke problem has been greatly reduced, and buildings and vegetation in industrial towns and cities are no longer covered in grime. Further improvements will certainly take place with the greater use of smokeless fuel, natural gas and nuclear power, but sulphur dioxide is still a major problem as this harmful gas is produced mainly by the burning of fossil fuels, especially in power stations. However, sulphur can be removed from fuel before burning, or retained after burning, although the process is costly.

Whenever combustion takes place oxygen is used up and carbon dioxide is produced. The normal recycling of carbon dioxide (p. 135) in the world ecosystem has kept the level of carbon dioxide in the atmosphere remarkably constant, but in recent years the vast increase in combustion has led to a 10% rise and it is still going up. Carbon dioxide in the atmosphere traps the sun's heat, and some scientists believe that by the year 2000 the world's average temperature may go up by 0·5°C. The danger of such a rise is that it could cause considerable melting of the world's ice, with a corresponding rise of ocean levels and consequent flooding. However, no actual rise in temperature has so far been discernible, due possibly to other counteracting factors such as the presence of atmospheric dust preventing some heat from reaching the earth. Not nearly enough is known about the consequences of these factors. Keeping the carbon dioxide cycle in mind, you will see that there is only one basic way of counteracting a rise in carbon dioxide levels—by extra photosynthesis. This is a factor to remember when considering the effects of increasing 'concrete jungle' at the expense of forests and green fields.

Motor vehicles are also major pollutants of the atmosphere as the combustion of motor fuel produces not only carbon dioxide, but also carbon monoxide, oxides of nitrogen and lead. This is so bad in some cities that in Tokyo, for example, traffic policemen are made to breathe pure oxygen after 20 minutes of duty. The carbon monoxide in exhaust fumes is very poisonous as it combines with the haemoglobin of the blood and so prevents it from carrying oxygen. The lead is also a poison as it accumulates in the body causing very serious illness. Research is going on to discover the most effective way of removing it. This is quite possible to do, but once more, it adds to the cost and somebody has to pay.

Nuclear fall-out is another form of pollution. After an atomic test, radioactive substances may reach the ground in rain, get taken up by plants, and may then pass up through the food web, becoming concentrated at each step. For example, the radio-isotope, strontium 90, can be taken up by cattle and passed on to man in milk and cheese, and end up in his bones. Here it may affect the bone marrow where blood corpuscles are made, causing **leukaemia** or blood cancer. Similarly, caesium

137 may be picked up directly from vegetables and concentrated in various organs of the body, including the gonads. Here it may damage the carriers of hereditary characteristics, the **genes** (p. 175), and so have a harmful effect on future children. Although the amounts of these radio-active isotopes found in children are very small, and not thought to be harmful, the danger lies in their cumulative effects. Strontium 90 is radio-active for a long time—it has a 'half life' of 28 years. This recognised danger from nuclear fall-out led to the Test-ban Treaty of 1963 which prevented further air testing of atomic weapons by Russia and America.

Water pollution

Rivers have always been the dumping ground for man's unwanted material. It is astonishing how much waste can be actually broken down into harmless products by bacteria, but there are limits beyond which rivers become stinking sewers. Nearly all aquatic life depends on the oxygen dissolved in the water and if too much organic matter is present the bacteria causing its decay will use up all the oxygen and other organisms will suffocate. It is then that the anaerobic bacteria take over and release evil smelling gases such as hydrogen sulphide.

We have already mentioned the problems of raw human sewage and how it can be processed before being passed into the water (Book 1 p. 193), the dangers of high concentrations of animal manure resulting from factory farming practice, and the run-off of artificial fertilisers which cause 'blooms' of algae to develop which may later die and decay. All these factors may help to make a river 'dead' through lack of oxygen. However, industry adds to the load of these **bio-degradable** substances—those which may be broken down by bacteria. In Japan, for example, the organic wastes from pulp and paper-making have rendered the rivers round the city of Fuiji completely dead. When large rivers empty their contents into lakes and inland seas, these in turn silt up rapidly, their oxygen content falls and they too become dead. Lake Erie in North America is a notorious example, but many lakes in Europe, and even seas such as the Baltic, approach this condition and support little life. To add to the problem there are the varied chemical effluents from factories and the accidental seepage of toxic agricultural chemicals into the river system. In three years it was estimated that 15 million fish were killed by the pesticide endrin which leaked into the Mississippi, and in 1969 when 90kg of endrosulphan fell off a Rhine barge in Germany, millions of fish died further down-stream, mainly in Holland. These are extreme examples of an event which, on a smaller scale, occurs commonly.

But what is being done about it? First, there has been a tightening of the laws in many countries regarding the dumping of raw sewage into rivers and the control of effluents from factories. Second, vast sums of money are now being spent on cleaning up rivers and lakes—especially in America and Europe. Some rivers, such as the Thames, heavily polluted 15 years ago, now have many species of fish in them once more. The situation is certainly improving in many countries.

What of the oceans? They are so vast that most people looked upon them, until recently, as an inexhaustible depository, but this is far from the truth. The oceans are the sinks into which effluents from the rivers pass. Pollutants come in, but they do not pass out! If they are bio-degradable, bacteria will act upon them, but if not they will accumulate.

Oil pollution is a particularly serious threat, not only because it fouls our beaches, but because of its effect on living things. A United Nations Committee estimated that over 2 million tonnes of oil finds its way into the oceans every year, about half of this by deliberate action. Collisions involving giant tankers and massive oil leaks from underwater drilling have already occurred, and in spite of the recognised danger, tankers become larger every year and underwater drillings are more common.

One obvious effect of oil on wildlife is the fouling of the feathers of seabirds, with their consequent death. It has been estimated that 10% of the eiderduck population in British waters died in this way when oil spilled out from an oil tanker after a collision in the Tay Estuary in 1968. When the giant oil tanker, the *Torrey Canyon*, ran aground off the Scilly Isles in 1967, the huge oil slicks were first tackled with the help of detergents, but these were found to be far more toxic to the wild life than the oil.

Today, more effective and less lethal methods are used, but prevention is always better than cure.

A fact of importance when considering the effect of pollution on the oceans is that so much depends upon the **phytoplankton**—those microscopic green plants in the surface waters of the sea which are the basis of nearly all oceanic food webs. Not only are they the main producers of the oceans, but they provide through photosynthesis a quarter of all the oxygen in the atmosphere. Anything that affects these vital plants adversely on a large scale is clearly a grave hazard. It is also significant that estuaries and shorelines are the main spawning grounds of fish and these areas are liable to be the most polluted.

The oceans have also been the dumping ground for the most dangerous poisons including stock-piled nerve gases and radio-active wastes. Although sealed in containers, leakages of some poisons have been known to occur causing large-scale deaths of fish and seabirds. A great and growing hazard arises in the disposal of nuclear waste which is increasing year by year as more atomic power stations are built. Most of this waste is either sealed in stainless steel tanks if it is liquid, or in other 'safe' containers if solid, and dumped in deep water. The operation is subject to the strictest control and there is regular inspection of the water and fish catches for signs of radio-activity, but the potential danger is very great.

Nuclear power stations need vast amounts of water to cool the reactors; this warmed-up water passes back into the rivers or sea causing **thermal pollution**. The rise in temperature greatly affects the organisms living there and may destroy many of them. However, in some places an attempt has been made to use this rise in temperature to good effect, as a few species such as eels and carp grow better in warmer water.

Land pollution

Man is producing more and more waste as industry and technology advances. Consider the effect of mining and quarrying. In the United States alone, 3 million tonnes of slag are being produced annually; this indicates the size of the problem of disposal. But a lot is being done; some waste is being left under-

Fig. 13:8 The effect of a selective weed-killer on a field of wheat: (left) untreated (right) treated.

ground in disused workings, dangerous tips in the vicinity of houses are being removed and others reclaimed by growing suitable plants which can withstand the toxic effects of many of these wastes. Much research has gone into finding the best plants to grow on these heaps according to their composition, in order to provide a gradual succession of vegetation and to build up a fertile soil once more.

Household and industrial waste is also increasing at an alarming rate. In addition to the garbage, food wrappings, plastic, metal and glass containers and waste paper that fill household bins every day, there is the increasing bulk of worn-out machinery, household furniture and unwanted cars to be disposed of. In the United States, to mention just a few items, 7 million cars, 48 billion metal cans and 26

Fig. 13:9 Combating soil erosion in hilly country by terracing: Maharashtra, India.

Fig. 13:10 Land destruction and the problem of waste: a colliery in South Wales.

billion bottles are discarded annually. All contain valuable material, but much of it is merely used for a time and then abandoned. Thus man is using more and more of the earth's resources and is wastefully throwing much of it away after use. We now realise that supplies of raw materials are not inexhaustible and that this kind of wastage cannot go on. So we are faced with two big problems: how to collect the more valuable material so that it can be recycled and how to dispose of the unusable residues without harming the environment.

Collection is the most difficult and expensive, because the waste is dispersed so widely; every house and factory has its quota, and even the roadside parking and picnic places are littered with cans, bottles and wrappings thoughtlessly thrown away by polluters of the countryside. In many countries the problem of litter is being tackled by the imposition of heavy fines, but this would not be necessary if each person took responsibility for his own litter.

Any method which separates bio-degrad-able waste such as garbage from other waste products is valuable, as the former can be treated and put back on the land as a useful fertiliser. In Holland over 30% of the waste from cities has been returned to the land as compost.

Disposal of waste is being tackled in many ways. These include:

1. Tipping

This is often the cheapest method and can be effective, especially in smaller communities, if suitable sites are available. The best way is for the refuse to be spread in layers about 2m thick and then covered with 30cm of soil before another layer is added. In this way it gradually builds up, like a layered cake, consolidates, and may eventually be reclaimed as useful land. Such tips, however, are apt to attract pests such as rats and flies and the use of pesticides to control them is usually advocated. In addition, their possible adverse effects have to be considered. Suitable sites for tipping are often hard to find and the habit of using wet-

160

lands can be strongly criticised as in many places these are valuable ecosystems in their own right for wildlife and may be useful fishing areas.

2. Burning

Modern incinerators have great advantages over tipping, which is at best a short-term policy. Cities such as Osaka, Paris and Düsseldorf have installed great plants which not only dispose of the rubbish, but may generate enough heat in the process to warm many of the buildings and also produce electricity. Metals are also recovered from the waste and the remaining ash is used for making concrete blocks for building.

3. Special processes

These include such machines as car-shredders, which crush and break up old cars so that the materials can be processed and used again.

The problem of land space

Man's history of land use has been one of haphazard development. Only in recent times has there been any real attempt to plan the best use for the available land. In many parts of the world much land has been spoiled, disfigured or wasted through lack of foresight, greed and expedience. But because of the pressure of expanding populations, massive technological advances and the necessity to grow more food, we are beginning to realise what a precious commodity land is.

Apart from the need to grow more food, there are many other uses to which land may be put. In the more densely populated countries, valuable agricultural land is being whittled away each year at an alarming rate as more houses are built and industries set up, more roads and motorways are constructed and more valleys flooded to meet the growing need for water. In the developing countries, with their fast expanding populations, some of the last of the major natural ecosystems are being destroyed as forests are cut down and scrub and hill-sides cleared to make way for agriculture. The vast areas of the African savannahs, for example, with their teeming wildlife are being invaded on all sides and the larger animals are being forced into smaller and smaller areas.

But 'man does not live by bread alone'. He needs land for other uses; more and more he needs relaxation from the pressures of modern life. He needs recreation; he needs to satisfy his deep-seated craving for natural beauty as an antidote for the artificiality of the urban environment. The wildlife with which he shares this planet also have their needs. Ecology makes it abundantly clear that we are part of nature, not apart from it. The world is a single ecosystem and there is a precarious balance between all the organisms which compose it. This balance has already been considerably disturbed by man. It is essential that natural ecosystems should remain and be studied so that he may understand the complexities of this balance and that his actions may not lead to further disasters. It is unthinkable that the magnificent heritage of wildlife that has taken hundreds of millions of years to evolve should, in the course of a few decades, be destroyed for ever. The need for national parks and smaller local nature reserves becomes more and more urgent as human activities expand. At present they take up less than 1% of the earth's surface and many of them, especially in America and Africa have become so popular as tourist attractions that human pressures on these parks are becoming a major problem. A compromise has to be reached between tourism and enjoyment, and the destruction by human pressure of the very ecosystem we wish to enjoy. This calls for careful management so that large areas are left undisturbed by the public. How then can so many conflicting claims on land-use be satisfied when land is already in such limited supply? It can only be done by the most careful planning at local, national and international levels, and by the scientific conservation of our whole environment.

The principle of **multiple land use** is an important one, particularly in regions of high population density. It can be illustrated by a scheme for the reclamation of a large area of peat moor in south-west England after the peat has been extracted for commercial purposes. The area is particularly valuable for its distinctive flora and fauna and in consequence some of it has been bought by the County Trust for Nature Conservation to prevent further peat extraction and to conserve the present ecosystem. But by far the greatest part will have its peat removed right down to the underlying clay. This will leave a vast 'hole in the ground'. The plan is to fill it with water and produce a valuable reservoir. Part of the lake thus formed will be kept as an amenity area for sailing, part will be colonised by reeds to become a wildfowl refuge and fish will be introduced for the anglers. In this way the one area can serve a variety of interests.

Conservation

In the widest sense this means the keeping of the environment in a state of balance; the restoration and maintenance of unpolluted air, land and water and the preservation by careful management of a great variety of ecosystems with their diverse communities of plants and animals. We have already referred to many examples of conservation. But we can only conserve our environment if we face the situation squarely, realistically and optimistically.

The world problems are serious. Man is in the grip of a population explosion without parallel in human history; unless this is stabilised the pressures will continue to increase. With his great technological advances he has replaced stable natural ecosystems with unstable artificial or semi-artificial ones. He has subdued nature, but in doing so has broken most of the ecological laws and brought an infinity of problems on himself in consequence. With the increasing demands of a consumer society he has dug deeply into the world's resources; he has taken out far more than he has put back for further use, and has very considerably polluted the environment on which he depends for life—how much, nobody yet knows.

But all is certainly not gloom! Man is capable of great wisdom as well as great destruction; he can analyse the difficulties he has created and reverse the dangerous policies he has set in motion. Already there is the knowledge and the technology to do this, and in this chapter we have seen how major problems are being solved. Many countries have launched effective programmes for pollution control and the recycling of the more important basic resources; rivers are being cleaned up, dangerous pesticides have been banned and less harmful alternatives used; land is being restored, erosion is being prevented, reafforestation is taking place, national parks and nature reserves are being created and managed. A start has been made, but there is a tremendous task ahead.

Conservation in its widest sense is of paramount importance; it is a matter of the greatest urgency involving governments and industry, science and technology, but above all it concerns each one of us. When we begin to understand the problems, are really concerned about them, and take responsibility for them in some positive way, if only on the smallest scale, we are on the way towards solving them. We all have to think and act ecologically—we have to live and work in harmony with each other and with the rest of the natural world, because the world is a single ecosystem which is the precious heritage and responsibility of us all. To quote from *Only One Earth* by Barbara Ward and René Dubos: 'Today in human society, we can perhaps hope to survive in all our prized diversity providing we can achieve an ultimate loyalty to our single, beautiful and vulnerable planet earth. Alone in space, alone in its life-supporting systems, powered by inconceivable energies, mediating them to us through the most delicate adjustments, wayward, unlikely, unpredictable, but nourishing, enlivening and enriching in the largest degree —is this not a precious home for all of us earthlings? Is it not worth our love? Does it not deserve all the inventiveness and courage and generosity of which we are capable to preserve it from degradation and destruction and, by so doing, to secure our own survival?'

For those who wish to become involved in the work of conservation here are some ideas:
1. Take part in such group projects as clearing footpaths and bridle ways, and restoring canals, ponds and rivers so that they may support thriving communities once more. If there is an ugly or derelict area in your neighbourhood, help to turn it into something more pleasant, such as a garden, a copse or a nature reserve.

For these projects you will need to obtain permission from owners and councils, and expert advice will be necessary. Consult such bodies as your District or County Council, the Nature Conservancy, the Forestry Commission and your County Trust for Nature Conservation.

Fig. 13:11 a) Oil pollution: a razorbill with oiled feathers. b) Egg of a peregrine falcon with thin shell due to a build-up of pesticides in the body of the mother. c) Fish killed by water pollution: Morocco, 1970. d) Wastes deposited and burned at Malmö, Sweden, causing air, water and land pollution.

Fig. 13:12 Members of the Conservation Corps re-claiming a pond in Britain.

2. Make your garden more attractive to wild life. Provide food and water for birds in winter and in times of drought. Erect nest boxes. If possible leave part of the garden 'wild' with plenty of cover. Grow plants which have a special attraction for bees and butterflies, and those on which the caterpillars of the butter-flies feed.

3. Join a local or national organisation devoted to conservation such as the County Trusts for Nature Conservation (most have conservation corps), the Wildlife Youth Service of the World Wildlife Fund or the British Trust for Conservation Volunteers.

14
Genetics

Variations

It is obvious that all members of a species have certain characteristics in common which distinguish them from other species, but within the species there is also much variation.

Looking round the class you will readily agree that you are all different; you may also agree, probably for the wrong reasons, that this is a very good thing! We vary in many ways; in the colour of our eyes, skin and hair, in the shape of our nose, whether or not our ears are lobed, in our height, weight and the proportions of our bodies, in our blood groups, our finger prints, in our ability to recognise colours and whether or not we can roll our tongue. The same applies to other animals; dogs and cats differ greatly from one another, so do the pigeons in our cities. It is more difficult to see variation, for example, in the members of a flock of gulls, but on close inspection differences can be detected. Certainly the birds themselves have no difficulty in recognising each other. Plants also show variation. Individuals of the same species may vary in height, in the number of petals, in the colour or shape of the seeds, in their reproductive capacity and so on.

Kinds of variation

You will see from some of the examples of variation given so far that they may be grouped into two categories. In the first, variations such as the ability to roll the tongue are quite definite and clear cut—you can either roll the tongue or you cannot. In the second, of which height is an example, there is a continuous range of intermediates between the two extremes. Consider the examples of human variation mentioned above; into which of these two categories would you place each?

Variations can also be classified according to whether they are inherited or not. It is easy to recognise that some clear-cut variations are inherited, but with those which show a continuous range it is more difficult, as environmental factors may also have an effect. For example, our general body shape and size is determined by heredity, but it is greatly modified by the amount of food that we eat and the exercise we take. The same applies to intelligence; heredity certainly plays a most important part, but the type of upbringing, especially during the early years, is a very significant factor too.

Mendel's experiments

Man has experimented with the breeding of plants and animals for many hundreds of years, in order to produce more useful varieties. He did so by selecting individuals for breeding which had the most useful characteristics. Sometimes the resulting progeny had the desired characters, but more often than not the opposite was the case. It was not until 1866 that **Gregor Mendel**, an Austrian monk who was both a biologist and a mathematician, published the results of his experiments which put the whole subject of inheritance on a sound scientific basis. The great significance of his work was not realised until 1900 when his paper, which had been published in a local journal, was 'rediscovered' by three eminent scientists working independently in different parts of Europe, and who continued the researches.

Mendel worked mainly on garden peas because they were easy to grow, had many distinct varieties and produced numerous seeds in a relatively short time. He noticed that some pea plants were tall, others short; some formed green seeds, others yellow; some had inflated pods, others constricted. He devised simple experiments involving the cross-breeding of varieties having these contrasting characters. Garden peas have flowers which are usually self-pollinated, but by removing the stamens early and dusting the stigmas

Fig. 14:1 No two cows are exactly alike.

Fig. 14:2 Children from the Seychelles: what variations can you detect?

with pollen from another variety, Mendel found that cross-pollination could easily be achieved. To make certain that no pollen was deposited from any other source, he covered the flowers with small paper bags.

Choosing plants which had bred true for a particular character for at least two generations, Mendel devised seven cross-breeding experiments, each involving one pair of contrasting characters. For example, tall plants were crossed with short, and plants with green seeds were crossed with plants having yellow seeds. When the seeds from these crosses were ripe he then grew them all in separate plots and noted which characteristic appeared in this **first filial** or **F$_1$** generation.

Fig. 14:3 Fruit flies (*Drosophila*): various mutants. a) Normal wings, apricot eyes. b) Vestigial wings, red eyes. c) Reduced wings, white eyes.

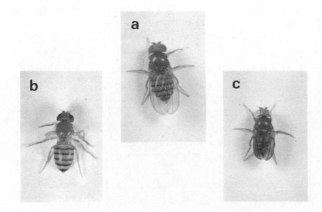

In every case *all* the progeny resembled *one* of the parents only. Thus when tall plants were crossed with short, *all* the progeny were tall. He then allowed all these F$_1$ plants to self-pollinate and again he carefully grew all the seeds that were produced. This time the results were very different. For example, when the tall F$_1$ plants were self-pollinated the resulting **second filial** or **F$_2$** generation consisted mainly of tall plants, but there were some short ones as well. He counted them carefully and worked out the ratio.

The results he obtained are shown in the table opposite.

Pairs of contrasting characters such as the above are now called **allelomorphic** pairs. The character which always showed in the F$_1$ Mendel called the **dominant**, the one that remained hidden, but appeared once more in the F$_2$ he called the **recessive**. But perhaps the most significant thing about his results was that the ratios all approximated to 3:1 in the F$_2$.

Mendel carried the experiments further by self-pollinating the F$_2$ plants to see what would happen in the F$_3$. For example, he found that all the short plants bred true, but of the tall ones about one-third bred true and two-thirds produced both tall and short plants in approximately the same proportion of 3:1 as in the F$_2$. Similar results were obtained for the other characteristics investigated. These results, using proportions only, are summarised in Fig. 14:4.

166

Experiment	Type of Cross	F₁ Generation	F₂ Generation	Ratio
1.	Tall (180–220cm)× short (22–45cm)	All tall	787 tall. 277 short.	2·84 :1
2.	Green seed× Yellow	All yellow	6022 yellow. 2001 green.	3·01 :1
3.	Round seed× Wrinkled	All round	5474 round. 1850 wrinkled.	2·96 :1
4.	Coloured seed coat× White	All coloured	705 coloured. 224 white.	3·15 :1
5.	Green pod× Yellow	All green	428 green. 152 yellow.	2·82 :1
6.	Inflated pod× constricted	All smooth	882 smooth. 299 constricted.	2·95 :1
7.	Flowers terminal× flowers axial	All axial	651 axial. 207 terminal.	3·14 :1

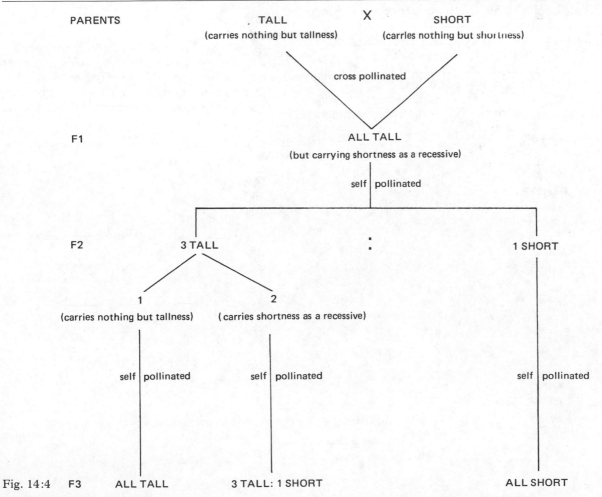

Fig. 14:4

167

How could these results be explained?

Mendel realised that there must be factors in pea plants which caused the characteristics such as tallness or shortness to develop. He had no idea what these factors were, so he used symbols. He called the factor for tallness T because it was dominant and the factor for shortness t because it was recessive.

Mendel also reasoned that these factors must occur in the parents as *pairs*, because in the F_1, although all the plants were tall they passed on the recessive character to the F_2, so there had to be a factor for tallness and another for shortness present. Consequently he called the parents TT and tt and the F_1 hybrids Tt.

Mendel also realised that if parent plants had pairs of factors, they must separate into single components when gametes were formed and pair up once more when fertilization took place, otherwise the numbers of such factors would double at each generation.

We now call these factors **genes** and so the study of inheritance is **genetics**. When parents have pairs of identical genes and therefore breed true, such as TT and tt, they are said to be **homozygous**, while those which are hybrids such as Tt, are **heterozygous**.

Using Mendel's symbols we can now express his results (Fig. 14:5), using a method which is now universally accepted and which you should use to work out any heredity problems.

Looking at the F_2 generation you should see at once why all the short plants will breed true when they are self pollinated, but only one third of the tall, as two thirds of the latter are heterozygous and will behave like the F_1 generation.

We thus have in the F_2, plants which *look* alike but *breed* differently; so we say that TT and Tt plants are of the same **phenotype** because they look alike, but as they carry different genes, they are said to be of different **genotypes**.

We can now summarise Mendel's conclusions using modern terms, as follows:

1. The genes which determine a particular characteristic occur in pairs (allelomorphic); they may be similar (homozygous) or different (heterozygous). When different, only the effect of one of them will be seen (the dominant), the other will remain hidden (the recessive).

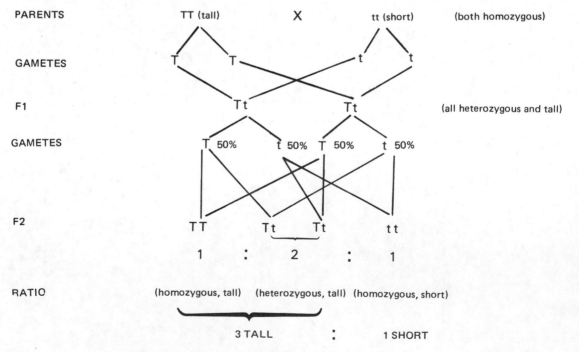

Fig. 14:5

168

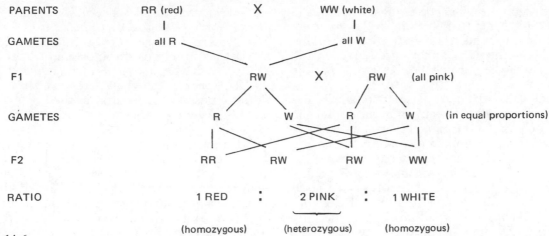

PARENTS   RR (red)   X   WW (white)

GAMETES   all R   all W

F_1 RW   X   RW   (all pink)

GAMETES   R   W   R   W   (in equal proportions)

F_2 RR   RW   RW   WW

RATIO   1 RED : 2 PINK : 1 WHITE

           (homozygous)   (heterozygous)   (homozygous)

Fig. 14:6

2. When sex cells are formed the genes of an allelomorphic pair become separated so that each gamete only carries one gene from that pair.

3. At fertilization the gametes fuse in a random manner, and in the case of those from heterozygous parents the ratio of the progeny approximates to 3:1 in favour of the dominant character. However, this is in fact a 1:2:1 ratio as one third of the dominant phenotype will be homozygous and two thirds heterozygous.

Referring to 1. above, there are now known to be instances where there is no marked dominance in the F_1. Consequently the heterozygous individuals show intermediate characteristics between those of the two homozygous parents. For example, some plants such as antirrhinums which have both white and red-flowered varieties, when crossed produce pink flowers in the F_1, but when these are self-pollinated or crossed between themselves the ratio in the F_2 is 1 red:2 pink:1 white (Fig. 14:6).

Are Mendel's ratios the result of random fertilization?

Mendel believed that the ratios he obtained in the F_2 were 1:2:1 because in the F_1 each heterozygous parent produced gametes of two kinds in equal proportions and it was entirely a matter of chance which fused with which. You can demonstrate this law of probability for yourself by doing some 'breeding' experiments using plastic beads.

Take two large beakers and place 100 red and 100 white beads in each. Each bead represents a gamete and each beaker will thus contain the two kinds of gametes produced by the F_1 in equal proportions. Let the beads in one beaker represent male gametes and the beads in the other the female gametes. Mix the beads thoroughly in each beaker and then extract one bead from each simultaneously without looking at them, i.e. at random, and place them together in pairs to represent the zygotes formed. Some pairs will be reds only, some white and some red and white. After you have taken out 12 pairs make a note of the ratio, repeat for 24 pairs and finally for 200 pairs. Compare your results with other members of the class. How nearly do your results approximate to 1:2:1? By using large numbers of pairs, were the results more, or less accurate?

You should now understand that, in practice, these theoretical Mendelian ratios are far more likely to be attained if the number of progeny is very large. Is this borne out by Mendel's seven experiments on peas? Are the most accurate ratios the ones where larger numbers were used? (Study the table on p. 167).

The mechanism of gene transfer

Mendel had brilliantly shown by his experiments that characteristics were inherited according to mathematical laws and it is now known that these laws apply to all plants and animals. But during his lifetime there was no explanation of what these genes were and what caused them to behave in the way they did. However, after his death it was discovered that the clue to the problem lay in cell division and that it was the chromosomes that carried the genes.

Each cell in the body of a plant or animal has a definite number of pairs of chromosomes in the nucleus, the number being characteristic of the species. For example, man has 23 pairs, a fruit fly 4 pairs and maize has 20 pairs. All through life when growth and repair is taking place the cells divide by **mitosis** (p. 62). During this process the number of chromosomes *remains constant*. This happens because each one duplicates itself to form two identical chromosomes, one going into one daughter nucleus and the second going into the other.

However, when reproduction takes place and gametes fuse, chromosome numbers would theoretically double, but in fact this does not occur because at some point in the life history the number is halved. This halving takes place in animals and some of the lower plants when gametes are formed, and in flowering plants during the formation of pollen grains. The type of cell division that brings this reduction about is called **meiosis**.

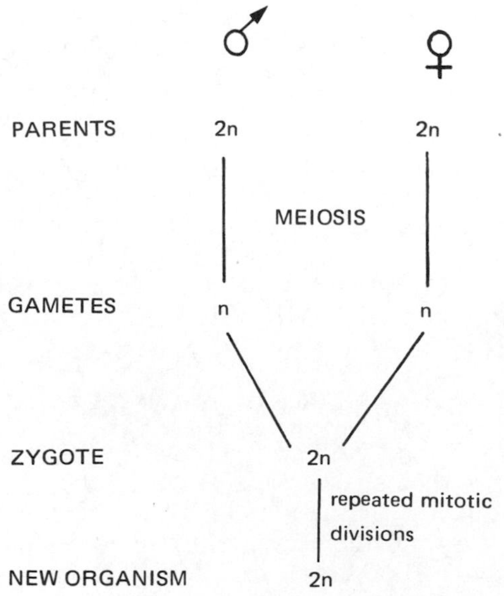

During meiosis the chromosome pairs become separated so that one *whole* chromosome from each pair passes into each daughter cell. Thus the number of chromosomes is reduced from 2n to n where n is a number characteristic for the species (Fig. 14:7). In man there are 46 chromosomes (23 pairs) in the body cells, but in the gametes there are 23 single chromosomes.

There is now much evidence that genes are carried by the chromosomes and it follows that because there are vastly more inherited characteristics than there are chromosomes, there must be a great many genes which are carried by each. It has been established that these are arranged in a linear manner like beads on a string and that allelomorphic pairs of genes such as those which control height, occupy corresponding positions on the two chromosomes which make a pair (Fig. 14:8).

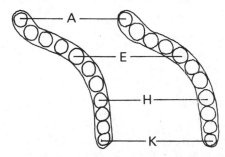

Fig. 14:8 Diagram illustrating the principle that allelomorphic pairs of genes occur in the corresponding position along the chromosomes.

If we now return to Mendel's experiments with tall and short peas we could depict his results using chromosomes instead of symbols (Fig. 14:9), but to simplify the diagram the number of chromosomes will be reduced to the one pair which carries this pair of genes. Let ● represent the gene for tallness and ○ the gene for shortness.

We can see therefore that the behaviour of chromosomes at meiosis and fertilization exactly confirms Mendel's hypothesis and explains how the genes become separated when gametes are formed and pair up once more in the zygote. It also demonstrates that in every chromosome pair one component has originally come from the father and one from the mother, hence there is an equal probability of father and mother contributing towards the characteristics of the resulting progeny.

Fig. 14:7

♂   ♀

PARENTS   2n   2n

MEIOSIS

GAMETES   n   n

ZYGOTE   2n

repeated mitotic divisions

NEW ORGANISM   2n

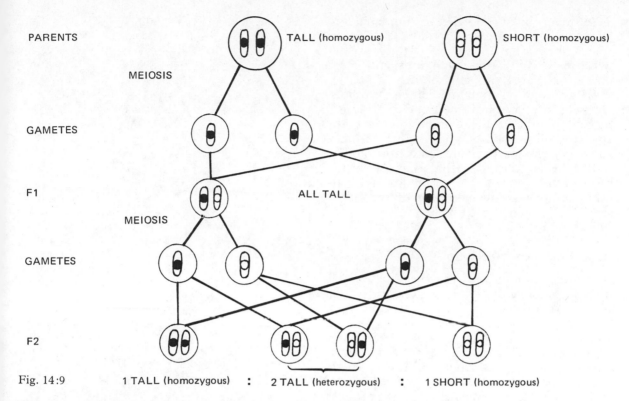

PARENTS   TALL (homozygous)   SHORT (homozygous)

MEIOSIS

GAMETES

F1   ALL TALL

MEIOSIS

GAMETES

F2

Fig. 14:9   1 TALL (homozygous)   :   2 TALL (heterozygous)   :   1 SHORT (homozygous)

Sex determination

In man and the majority of animals the sex ratio of the progeny is approximately equal. This is because one pair of chromosomes is concerned with the determination of sex. In the majority of animals, in the female, this pair consists of chromosomes which are similar in size and shape and are called X chromosomes, but in the male one is a normal X chromosome and the other is much smaller and is called the Y. When the female forms gametes and the chromosomes separate in meiosis each egg will contain an X chromosome, but in the male, when sperms are formed, 50% will contain X and 50% Y. So it follows that with random fertilization there is an equal probability of a male or female zygote being formed.

Fig. 14:11 (below) Photograph of human chromosomes (male) from a body cell. There are 46 altogether composed of 23 allelomorphic pairs. Each pair is similar in size and shape except for the X and Y.

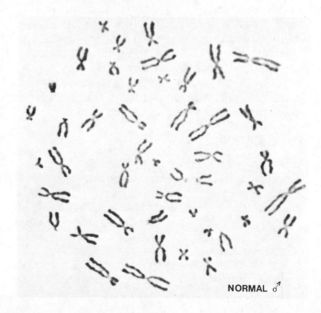

NORMAL ♂

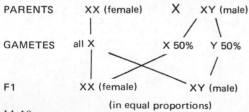

PARENTS   XX (female)   X   XY (male)

GAMETES   all X   X 50%   Y 50%

F1   XX (female)   XY (male)

(in equal proportions)

Fig. 14:10

In man this proportion is not achieved exactly as the probability of a Y-carrying sperm fertilizing an egg is slightly greater because it is fractionally lighter and more active than the X-carrying sperm.

It is not always the male that carries the Y chromosome; in some animals, e.g. birds and insects, it is the female, hence in these it is the egg which either contains an X or a Y chromosome, and the sperms which all contain an X.

If follows from this that the sex of the progeny is determined at fertilization, although the later development of secondary sexual characters is influenced by the sex hormones (p. 33).

Breeding experiments

It is interesting to carry out breeding experiments yourself, but to be successful over a limited period of time it is essential to choose a species which breeds rapidly, is easy to keep and produces a large number of progeny. For these reasons the fruit fly *Drosophila melanogaster* is a suitable choice. Much genetic research has been done on this species.

The wild type has large wings, red eyes and a striped abdomen, but many genetic varieties occur (Fig. 14:3) which have contrasting characters such as vestigial wings (short and shrivelled), white eyes or ebony body (dark body with no stripes). Any of these varieties may be crossed with the wild type.

The sexes may be distinguished quite easily (Fig. 14:12). The female is longer and has a more pointed abdomen and is not so dark at the tip. When examined under a binocular microscope the sex comb may be seen on the foreleg of the male, and there are obvious differences in the genitalia at the tip of the abdomen when viewed ventrally.

Breeding the flies

They may be bred in wide mouthed bottles or large specimen tubes using a specially prepared culture medium as a food supply. A folded piece of paper towelling is placed inside (Fig. 14:13). When flies are introduced, they lay eggs on the medium and the semi-transparent larvae which soon hatch out tunnel into the food. When they are fully grown they climb up the towelling to pupate. The flies hatch out in about 12 days if kept at 25°C.

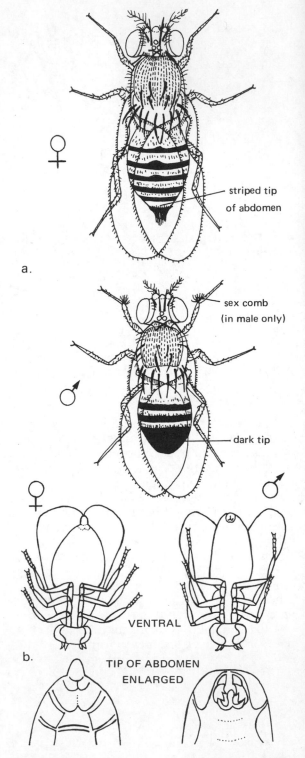

Fig. 14:12 Sex differences in *Drosophila*: a) dorsal b) ventral.

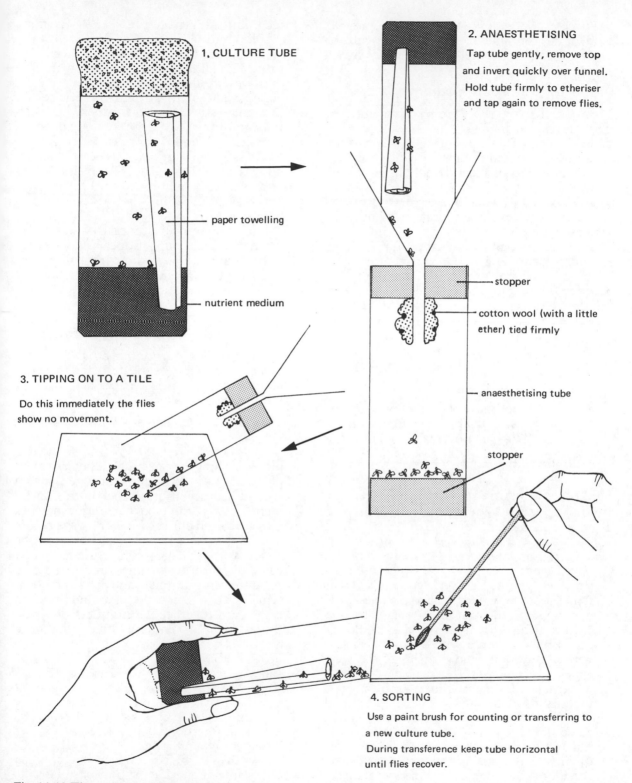

1. CULTURE TUBE

paper towelling

nutrient medium

2. ANAESTHETISING

Tap tube gently, remove top
and invert quickly over funnel.
Hold tube firmly to etheriser
and tap again to remove flies.

stopper

cotton wool (with a little
ether) tied firmly

anaesthetising tube

stopper

3. TIPPING ON TO A TILE

Do this immediately the flies
show no movement.

4. SORTING

Use a paint brush for counting or transferring to
a new culture tube.
During transference keep tube horizontal
until flies recover.

Fig. 14:13 The technique for handling *Drosophila*.

173

Successful breeding depends on keeping all food and containers sterile. The culture medium may be prepared in the following way:

Preparing the food medium

Enough for 60 100 × 25mm specimen tubes may be made as follows:
1. Soak 70g of fine oatmeal in 120cm³ of water for several hours.
2. Dissolve 30g of black treacle in 40cm³ of water.
3. Add 6g of powdered agar to 400cm³ of water, stir and then heat to boiling to dissolve it.
4. Boil all the above together for 15 min adding 6cm³ of a 10% solution of nipagin in 95% alcohol—this is an anti-mould.
5. Pour the medium into the clean specimen tubes to a depth of 25mm while still hot.
6. Sterilise the tubes and their plugs by keeping them under pressure in an autoclave for 15 minutes. Cool.
7. Push into the food, while still warm, a piece of folded paper towelling or filter paper which has been sterilised.
8. When the food has set add a few drops of yeast suspension and plug the tubes with the stoppers. They are now ready for culturing the flies.

Making experimental crosses involving wild type and ebony body

1. *To find out which is the dominant, normal body (wild type) or ebony body.* Given pure breeding stocks of these two varieties, you will need to separate 5 virgin females of the wild type and 5 males with ebony body and put them together in a culture tube. Others in the class could do the reciprocal cross with 5 virgin females with ebony body and 5 males of the wild type.

Female flies do not mate until at least 8 hours after hatching so in order to obtain virgin females it is necessary to remove all adult flies from a culture which is hatching and then use the females which subsequently hatch during the next 8 hours. As the flies are active it is necessary to anaesthetise them lightly with ether using the technique shown in Fig. 14:13. Examine the flies on the tile under a lens, select the ones you need and place these in a culture tube. Keep the tube horizontal when the flies are put in and wait until they recover before standing it upright, otherwise the flies may stick to the medium and die. Repeat for the second stock. You now have 5 males (♂) and 5 females (♀) together. Label the tube, e.g. ♀ wild type × ♂ ebony body and add the date. Keep at 25°C if possible. They will breed at temperatures lower than this, but they take longer to hatch.

Examine your cultures each day, and note when larvae and pupae appear, and the date of hatching. Remove the parent flies after 3 days. Why? When the flies hatch, find out which characteristic is dominant.

2. You should now carry out one or preferably both of the following different crosses:

a) *Breed the F_1 among themselves*
Select 5 of each sex and transfer to a new culture tube. It is not necessary to select virgin females. Why?

b) *Make a back cross with the recessive parent*
Select 5 virgin females of the F_1 and 5 males from the original ebony-bodied parental stock and put them together in a second culture tube. Label both experiments. When the next generation hatches out leave each culture for 2 or 3 days before counting as it takes several days for the majority to emerge. Then etherise each colony in turn, tip the flies on to a tile, and separate into normal and ebony-bodied types. What ratios do you find in each case? Add together the numbers obtained by all the class and find the ratios once more.

Can you explain why there is a difference between the results of a) and b) above? You can find the theoretical proportions by calling the original parents NN for normal type and nn for ebony body. The two crosses you made involving the F_1 were therefore a) Nn × Nn and b) Nn × nn. What progeny would result and in what proportions? How do the proportions compare with your results?

What are genes?

It is probable that most organisms have more than a thousand genes in their chromosomes. Each gene is composed of DNA. This substance has a remarkable molecule of very great size and complexity which occurs in an infinite variety of forms. Genes are portions of this giant molecule. In 1953, the detailed structure of DNA was finally worked out at Cambridge by **Francis Crick** and the American, **James Watson**, in close collaboration with **Maurice Wilkins** of King's College, London. For this achievement they were jointly awarded a Nobel prize. They showed that this DNA complex, which is thought to run down the chromosome, looks like a spiral staircase; the framework consists of alternating sugar and phosphate groups and the steps which join the framework together are pairs of chemical compounds called bases. The shape of this molecule is known as a double helix (Fig. 14:14). These bases are of four kinds, known as A, T, C and G,

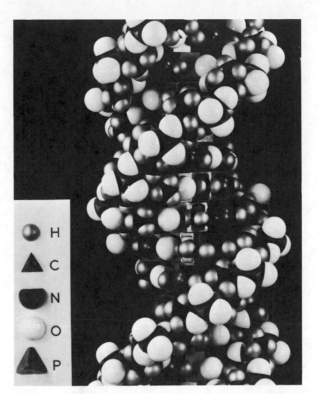

Fig. 14:15 Photograph of a model of part of the DNA molecule.

which unite with each other in a special way, A with T, T with A, C with G, and G with C to form each step. The order of bases arranged on the DNA molecule is almost infinitely variable and it is this variation that provides the genetic code which determines the characteristics of any individual. What is more, these long spirals of DNA are capable of replicating themselves. During mitosis they split exactly down the centre of the 'staircase' and each half step picks up another unit complementary to itself from free material in the nucleus; thus a new double helix is formed (Fig. 14:16). In this way exact replicas of the genes are passed on from cell to cell at each division until they become part of the nucleus of a gamete which can pass them on to the next generation. The genetic code enables chemical information within the genes to be translated so that the cytoplasm builds up the correct proteins within the cells of the body. These proteins, many of which are enzymes, are responsible for the metabolic processes which take place in the cells and so help to determine the particular characteristics of the individual.

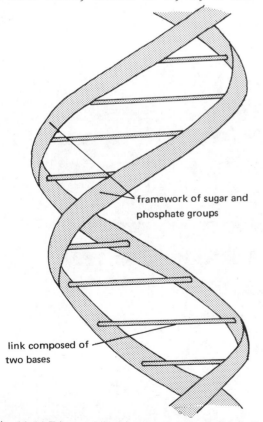

framework of sugar and phosphate groups

link composed of two bases

Fig. 14:14 Diagram showing part of the double helix of the DNA molecule.

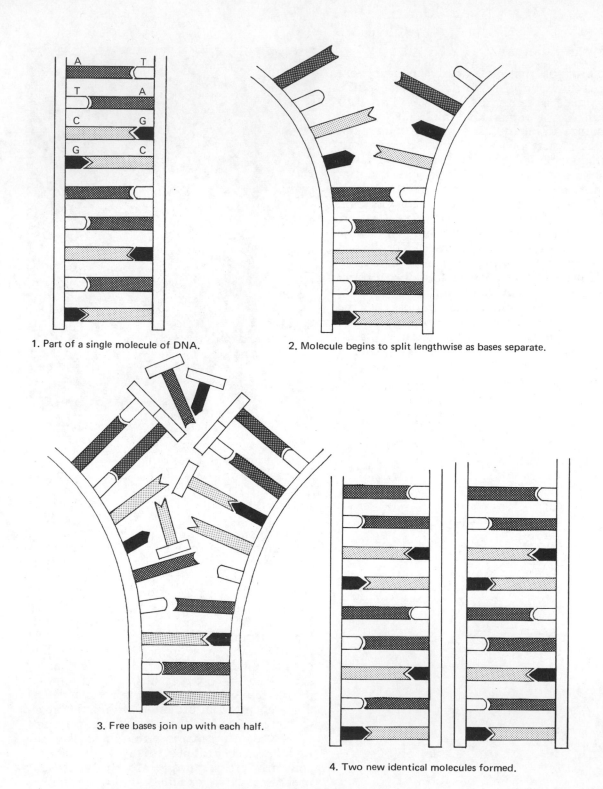

1. Part of a single molecule of DNA.

2. Molecule begins to split lengthwise as bases separate.

3. Free bases join up with each half.

4. Two new identical molecules formed.

Fig. 14:16 Diagram illustrating the replication of the DNA molecule.

The genetic basis of variation

Whenever sexual reproduction takes place new combinations of genes take place which result in a vast number of small variations in the progeny. The process has been picturesquely described as 'reshuffling the DNA pack of cards'. However, this method merely makes use of genes already present in different combinations, it does not produce entirely new ones.

Mutations

At infrequent intervals completely new genes are formed, called **mutants**. It is believed that mutation occurs as a result of some accident during the replication of the DNA molecule. If a mutation occurs in a body cell during mitosis, all the cells derived from it will have the same new characteristic, but as gametes are not involved, it will not be passed on to the progeny. These are called **somatic mutations** and produce such conditions as cells without chlorophyll in variegated plants. If, however, the mutation occurs in a reproductive cell, it will be transmitted if the gamete takes part in fertilization, and the new individual will pass it on to future generations. Usually, such a gene mutation causes only a minor change such as red eye colour to white in *Drosophila*, but major changes occur occasionally. These major mutations are always harmful and often cause the death of the progeny at an early stage of development; they are therefore called **lethal** genes.

It is not known what causes a mutation, but under natural conditions they occur very infrequently, sometimes no more than once in a million cell divisions. The likelihood of mutation is increased by some environmental factors such as temperature and also by various forms of radiation. It is possible to produce mutations artificially by subjecting the reproductive cells of an organism to X-rays. The first person to do this was **Hermann Muller**, an American geneticist, who in 1927 experimented on *Drosophila* and produced all sorts of mutants including some with various eye colours and others with abnormally shaped wings and appendages. It is because X-rays may affect genes that it is wise not to subject the gonads to X-ray treatment. For the human race the greatest potential mutation hazard comes from atomic radiation. The higher the level of radiation in the environment the more likely it is for mutations to arise. Nuclear warfare is the ultimate hazard in this respect.

The majority of natural mutations cause very small changes; some may be beneficial, others harmful; the majority are recessive. This means that they may not become apparent for a few generations. Only when the gene has spread within the population is it likely to combine at fertilization to form a double recessive and so become visible.

Chromosome re-arrangement and multiplication

Mutations do not always result from the change in a gene. Sometimes they occur when whole chromosomes behave abnormally. Two examples will be given.

1. Occasionally during meiosis, when gametes are formed, a complete chromosome pair passes into one gamete and none into the other. If the former then fuses with a normal gamete there will be three chromosomes of one kind instead of two. This sometimes happens in the 21st chromosome pair in man, producing in consequence a mentally-handicapped child known as a **mongol**. This abnormality becomes more probable with the increasing age of the mother; thus the risk of having a mongol baby is 1 in 2000 at the age of 20, but increases to 1 in 50 at 40.

2. Polyploidy. This is a phenomenon found in plants where the number of chromosomes is a multiple of the number normally found in the gametes, i.e. 3n, 4n, 6n, etc. Many garden varieties of flowers are polyploids and so are many modern kinds of fruits and cereals. Polyploids are often associated with an increase in size or number of parts such as petals.

15

Organic evolution

Fig. 15:1 Charles Darwin (1809–82).

The theory of evolution postulates that all organisms living today have been derived from pre-existing forms by a series of changes throughout immense periods of time; that all living things have a common ancestry and, starting from simple beginnings, there has been a gradual progression towards increasingly complex forms of life.

The idea of evolution is a very ancient one going back to the time of the Greeks, but it was not taken very seriously until the 19th century when **Charles Darwin** (1809–82) published his famous theory about it. Charles Darwin will go down in history as one of the greatest English naturalists. Many of his ideas about evolution stemmed from his five-year voyage round the world in the *Beagle*—a British naval ship carrying out survey work. Darwin joined the ship as naturalist to the expedition. Darwin had an enquiring mind and displayed great powers of critical observation; he also paid meticulous attention to detail. On his return to England he patiently and methodically assembled his evidence and continued his researches. By an extraordinary coincidence, another naturalist and explorer, **Alfred Russell Wallace** (1823–1913), when working in south-east Asia had come to similar conclusions to those of Darwin although neither was aware of the other's work. Wallace actually sent Darwin an account of his theory, asking him for his opinion about it and whether it should be published! Eventually it was de-

cided that they should jointly put forward their theory at a meeting of the Linnaean Society in London. Darwin's book, *The Origin of Species by means of Natural Selection* was published the year after, in 1859. This book not only provided a considerable body of evidence in support of evolution but it also went a long way to explain the means by which it could have been accomplished.

The theory of natural selection

When Darwin propounded his theory he had no knowledge of chromosomes, genes or mutations and therefore had no idea how variations were inherited; all this came later, but otherwise the modern theory is largely based on his conclusions. The theory may be summarised as follows:

1. In every species the individuals vary greatly among themselves; some of these variations are acquired during the life of the individual as a result of environmental factors—these

are *not* inherited and hence are of *no* significance in evolution. However, changes involving genes and chromosomes *are* inherited and these *can* lead to evolution.

2. In most species the number of progeny is far greater than is necessary to replace the parents when they die. Many in fact are so prolific that if there were no check to their numbers they would soon be unable to acquire the necessities for life. In consequence there must be competition among members of the same species for all these necessities. This is what Darwin called 'the struggle for existence'.

3. As a result of this competition, those individuals with variations which gave some advantage in this struggle for survival would be more likely to survive and breed. Darwin described this as 'the survival of the fittest', and he reasoned that it was nature (meaning all the environmental factors, physical and biotic) which did the selecting of the fittest individuals; so he called the process 'natural selection'.

4. As a consequence of natural selection occurring generation after generation over immense periods of time, populations would gradually change and new species evolve.

An example of natural selection today

One of the best documented examples of natural selection concerns the appearance as a mutation of a black variety of the peppered moth, *Biston betularia* in the Manchester district towards the middle of the last century. The normal form of this moth is speckled and much lighter in colour. The mutant resulted from a change in a single gene, and rather exceptionally it happened to be a dominant. The mutant type quickly spread, and in less than a century had almost completely replaced the normal type in many parts of Britain. In the 1950's a countrywide project was mounted to work out the distribution of the two forms and it was found that the black mutant was most common in industrial regions where, because of air pollution, the bark of trees and the sides of buildings on which the moths normally rested during the day were dark with soot deposits. The light variety was more common in unpolluted areas. It seemed a likely theory that birds were selectively eliminating the light ones in the polluted areas, because they were more conspicuous against the dark background (Fig. 15:2). To test this idea, large numbers of both varieties were bred and released in both polluted and unpolluted areas

Fig. 15:2 The peppered moth (*Biston betularia*) and its black mutant: (left) both kinds at rest on a soot-covered oak trunk near Birmingham; (right) both kinds at rest on a lichened tree trunk in unpolluted countryside.

and after a short period, as many as possible of those which survived the predation of birds were attracted to mercury vapour moth traps and counted. The results were as follows:

| | Released | | Recaptured | |
	Black	Light	Black	Light
Polluted area (Birmingham)	477	137	40%	19%
Non-polluted area (Dorset)	473	496	6%	12·5%

A film was also taken which showed the birds actually searching the tree trunks; this confirmed that they selected in each case more of the variety which was most easily seen.

An interesting sequel to the story is that since the Clean Air Act of 1956 pollution has gradually become reduced and now the light varieties are once more becoming common near industrial cities.

This is an example of a mutation which provided an advantageous characteristic in a polluted area, its survival depending on the selective effect of the environment, in this case birds. But it is important to realise that the same mutation when it occurred in an unpolluted area, was a disadvantage. So the principle is that natural selection of the fittest types concerns only those individuals which survive at a definite time in a particular place. It follows therefore that when environmental conditions remain fairly constant over long periods, the organisms living there will gradually become more perfectly adapted to those conditions due to natural selection. So if mutations occur under these circumstances, the likelihood of their being advantageous is not great as the organisms are already so well adapted. However, such a mutation, if recessive, will nevertheless become part of what is called the **gene pool**, i.e. the total number of genes present in a given population. It may therefore appear from time to time as a double recessive and if by that time environmental conditions have changed, the mutation may possibly be advantageous under these new conditions and consequently it may spread through the population.

To summarise, we can say that mutation and gene shuffling during meiosis produce the variations, mendelian inheritance governs their transfer from one generation to the next, and the environment brings about the selection of individuals which, on balance, have the most advantageous combination of genes. So although mutations occur in a random manner, selection ensures that evolution is directional.

How do new species evolve?

We saw with the peppered moth how a mutant gene could bring advantage to a species and cause a change in the population, but this is a long way from becoming a new species. For this to happen a population has first to be isolated from other populations in some way, so that inter-breeding is avoided or becomes less likely. There are many geographical barriers both large and small which could separate populations, such as land barriers which divide aquatic populations from each other, water barriers which divide terrestrial populations and mountain ranges or deserts which have a similar effect.

Let us assume that two populations have been isolated by some geographical barrier. We can postulate that mutations will occur in each, and natural selection will take place according to the environmental conditions prevailing in each region. As these conditions will be different in the two regions, the differences between the two populations will gradually become greater under the influence of natural selection until they are sufficiently different to prevent inter-breeding and the production of fertile offspring. In this way a new species is evolved. That is the theory; but what evidence is there in its favour?

Islands

These are excellent examples of isolation, so it is logical to assume that if they have been islands for a very long time they will probably contain species found nowhere else. This is exactly what is found! Perhaps the most famous examples are the Galapagos islands, 1000km off the west coast of Ecuador in the Pacific, and the Seychelles and Aldabra 1400km from the African mainland in the Indian Ocean. All these islands have many species found nowhere else in the world.

The Galapagos islands probably arose as a result of volcanic action and gradually became populated by a flora and fauna which reached them by flying, swimming and being carried by ocean currents or by the wind. In this way they became isolated from the populations on the mainland. Today many of the animals that live there still resemble the mainland forms in many ways, but they differ markedly in others. Thus there are flightless cormorants which resemble the Chilean species except for the reduction of the wings. Others, such as the giant tortoises and the iguanas, resemble species which have long since become extinct elsewhere, but have survived on the islands because of lack of competition or predation.

Additional evidence for the evolution of new species occurs within the islands themselves. For example, the giant tortoises occur in several of these islands and although the populations are not separated from each other by great distances, they nevertheless show distinct differences between them. But the most spectacular example is that of the finch-like birds. There are 13 species of these which differ markedly from each other, although they almost certainly have evolved from a single species. They differ particularly in the shape of the beak, which is adapted in different species, either for feeding on various kinds of seeds, or on cacti or on insects (Fig. 15:3), but they also vary in size and behaviour. Some species are found singly on the more remote islands while others occur together on the same island, but occupy different niches in the island ecosystem. So islands provide remarkable evidence that evolution has occurred.

VEGETARIAN TREE FINCH

WARBLER FINCH
(insects)

WOODPECKER FINCH
(insects)

LARGE INSECTIVOROUS
TREE FINCH

CACTUS GROUND
FINCH

LARGE GROUND FINCH
(seeds and fruit)

Fig. 15:3 6 of the 13 species of finch-like birds found on the Galapagos Islands. They are believed to have evolved from a single species as a result of isolation. The differences in beak size and shape are adaptations related to variation in their diet.

Comparative anatomy as evidence for evolution

The theory of evolution explains a mass of facts which are difficult to explain in any other way. Some of these concern the structure of organisms. Consider these questions and whether evolution explains them.

1. Why is it that plants and animals can be arranged in a progression from simple forms to complex?
2. Why should a phylum or class be built on a common plan of structure, the same structures being used for very different purposes? Take, as examples, the forelimbs of vertebrates used for running, climbing, swimming and flying (Book 1 p. 113), or the mouthparts of insects (Book 1 p. 69).
3. Why are **vestiges** present in animals? Vestiges are structures which are useless to the possessor, but correspond to useful organs in other animals. They include structures such as the remains of hind limbs in whales and pythons, the remnants of wings in some moths and the splint bones in the limbs of horses (Figs. 15:4, 15:15).

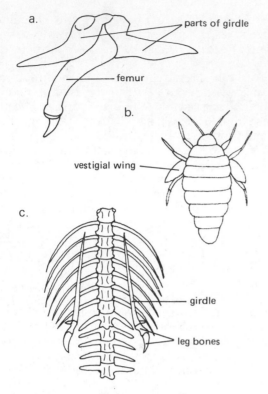

Fig. 15:4 Some examples of vestiges: a) hind limb of a whale (buried in blubber under the skin); b) reduced wings in the female vapourer moth; c) hind limb of a python which projects as a claw.

4. Why is it that some animals show intermediate characters between different classes of animals? Take, for example, the lung fish which have characteristics of both fish and amphibia, and the platypus which shows some intermediate characters between those of reptiles and mammals.

Do you agree that by assuming a common ancestry and gradual change these questions could be logically explained?

The fossil record of evolution

It is estimated that the earth may be something like 5000 million years old and that conditions were such that life could have come into being between 3 and $3\frac{1}{2}$ thousand million years ago. It probably took another 1 or 2 thousand million years before living things became cellular. The oldest fossils known are those of bacteria and blue-green algae which are estimated at over 1600 million years old. They have been found beautifully preserved in silica.

The thrilling story of the progression of life throughout the ages is written for us in the earth's rocks. The older parts of the story are very fragmented and large parts are missing, but nevertheless in the sedimentary rocks of the past 600 million years the fossils tell a remarkable story. These sedimentary rocks were laid down as sediments—depositions of materials formed as a result of erosion by water, wind or ice. As other layers formed above them they became more and more compressed to form the sandstones, limestones and shales that we know today.

Fossils

Normally when organisms die their bodies quickly decay and no traces are left, but if they have hard structures such as shells, bones or scales they may persist longer. So, for fossilisation to take place an organism must quickly be protected from decaying action and from the effect of weathering, and this is a comparatively rare event. Usually it occurs when organisms become quickly buried in mud or sand, or less often, in volcanic laval flows or the resin which exudes from trees. Insect fossils, for example, are commonly found in amber, which is fossilised resin. Some of the most abundant fossils are the Foraminifera (Book 1 p. 30)—those tiny Protozoa whose calcareous shells settled at the bottom of shallow seas to form an ooze which later consolidated into the chalk deposits we know today. Of the larger species, the calcareous shells of molluscs are often well preserved and may be abundant, for example, in certain limestones.

Fig. 15:5 Ammonites—the fossilised coiled shells of extinct molluscs.

Fig. 15:6 A marine iguana from the Galapagos Islands: a product of isolation.

Fig. 15:7 Giant tortoise from the island of Aldabra—its last stronghold in the Indian Ocean.

Fig. 15:9 Duck-billed platypus: a mammal which lays eggs.

Fig. 15:8 African lung fish.

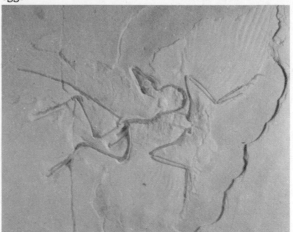

Fig. 15:10 Fossil of *Archaeopteryx*—the earliest known bird.

The Palaeozoic era (from about 600–225 million years ago)

The main steps in the progression from simpler organisms to the more complex are shown in Fig. 15:11. Fossil remains from Pre-Cambrian times are scarce possibly because at that time few animals had hard parts which would easily fossilise. But there is abundant evidence that in the Cambrian period at the beginning of the Palaeozoic era a great diversity of animals had already evolved. At this time the land seems to have been devoid of life, but in the shallow seas all the major invertebrate phyla were already represented (Fig. 15:12).

The first vertebrate fossils came from the Ordovician period—strange fish-like creatures with round mouths, remote ancestors of the lampreys of today. It was not until the Silurian period that true fish with jaws are thought to have made their appearance. However during the Devonian they became the dominant class of creatures in both sea and fresh water and showed great variety of structure. Both bony fish and cartilaginous forms such as sharks occurred at that time.

It was in the Silurian period that the land began to be colonised. Up to this time the only green plants were aquatic algae and no fossils have been found of the earliest land plants. Presumably these were flat seaweed-like algae which could withstand some degree of desiccation. However, by the middle of the Silurian period plants, called psilopsids, were present which lived under marshy conditions and had erect stems with spore-bearing structures at their tips. In the Devonian many other kinds of land plants developed which were the ancestors of our horsetails, club mosses and ferns. These had well developed lignified vascular systems that allowed them to grow to a considerable size. This led to the production of swamp forest with giant horsetails and club mosses which reached a height of 30m in the Carboniferous period. The remains of the vegetation from these swamp forests, consolidated and hardened, formed the coal measures of today. This luxuriant flora provided much food for the animals which, like the plants, were gradually invading the land. These included land snails and primitive insects with biting mouthparts such as cockroaches and dragonflies. Spiders, scorpions and centipedes were also present, and these presumably fed on the vegetarian insects.

The first vertebrates to come on land were primitive amphibia. They were still very dependent upon wet conditions and closely resembled the lobe-finned fishes from which they had evolved, except that their paired fins were replaced by limbs and their lungs had developed sufficiently for land respiration. Towards the end of the era, i.e. in the Permian period, there appeared the first vertebrates to be independent of water for reproduction. These were the first reptiles. The era ended with great upheavals of the earth's crust all over the world; inland seas were drained of their water and continents became uplifted.

The Mesozoic Era (from about 225–65 million years ago)

As the land became higher and drier the swamp forests of the last era were replaced by forests of conifers, and by the early Cretaceous period some angiosperms had certainly evolved. Before the era ended there was a profusion of flowering plants, including some which closely resembled our modern genera. This evolution of many kinds of flowers was closely paralleled by that of the insects. As the flowers became tubular and the nectar more difficult to reach, so the mouth parts of the insects became longer. Flies and bees were numerous at this time.

The Mesozoic era, however, is primarily known as the age of reptiles, because during this period they became the dominant vertebrate group. They reached their peak both in size and diversity in the Jurassic period when they colonised all the main habitats. Some, such as the fish-shaped ichthyosaurs returned to the water; others called pterosaurs took to the air with wings more akin to those of bats than birds with a membrane stretching from their elongated fore-limbs to the hind part of the body. Others lived an amphibious existence resembling in many ways the crocodiles of today, while terrestrial forms diversified in many directions. The most spectacular reptiles were the dinosaurs, some of which reached a length of 25m and a weight of over 30 tonnes. Not all the dinosaurs were large, but the biggest must have lived an amphibious existence, using the water to reduce the enormous weight on their

PLANT EVOLUTION	DURA-TION	ERA	PERIOD	YEARS AGO	ANIMAL EVOLUTION
	1–2	CENOZOIC	PLEISTOCENE	2	
	5		PLIOCENE	7	
	19		MIOCENE	26	
	12		OLIGOCENE	38	Age of mammals and birds
	16		EOCENE	54	
Age of flowering plants	11		PALAEOCENE	65	
	71	MESOZOIC	CRETACEOUS	136	
	57		JURASSIC	193	Age of reptiles
Age of gymnosperms	32		TRIASSIC	225	
	55	PALAEOZOIC	PERMIAN	280	Age of amphibians 
 Age of ferns and horsetails	65		CARBONIFEROUS	345	
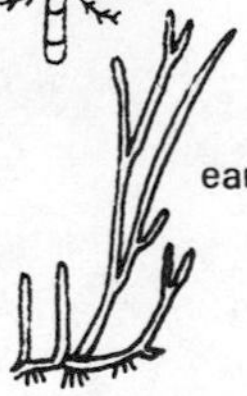	50		DEVONIAN	395	Age of fishes 
Age of early land plants	40		SILURIAN	435	
	65		ORDOVICIAN	500	Age of invertebrates 
Age of algae	100		CAMBRIAN	600	
Few fossils	millions of years		PRECAMBRIAN	millions of years	Few fossils

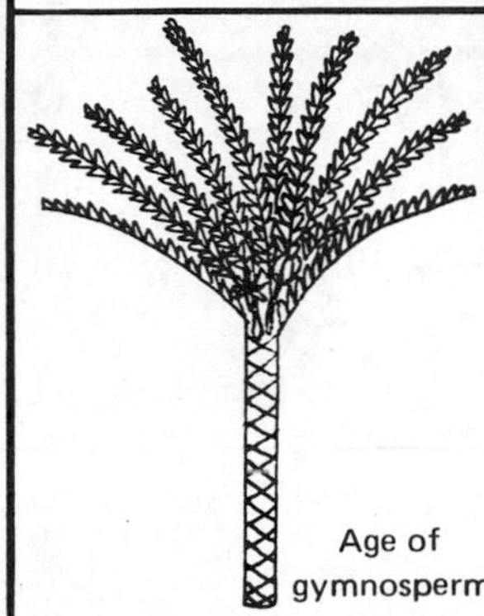

Fig. 15:11

Fig. 15:12 An artist's impression of life in the Palaeozoic Era.

Fig. 15:13 Models of dinosaurs from the Mesozoic Era.

Fig. 15:14 An artist's impression of some of the mammals of the late Cenozoic Era.

legs. Their teeth indicate that they fed on lush vegetation. There were also huge carnivores present which preyed on the herbivorous dinosaurs. They reached a length of 15m, weighed twice as much as a large elephant and had jaws of enormous strength armed with pointed teeth. They ran on two legs, their massive tails helping them to keep their balance.

During this era of reptilian evolution some were evolving towards birds and mammals. The earliest fossil evidence of a true bird was that of *Archaeopteryx* in the Jurassic period; it showed remarkable characters intermediate between those of a lizard and a bird. It had a bird-like beak, but rows of teeth were present as in reptiles. It had a long lizard-like tail, but with a double row of feathers attached to it. It also had feathers on its wings, but three clawed fingers were present which it probably used for clambering. This fossil alone provides powerful evidence in support of evolution.

It was not until the Cretaceous period that birds became numerous and evolved into types which resemble some of the more primitive groups we know today, such as pelicans, flamingoes and herons.

Mammals also evolved from reptiles during the Mesozoic era and by the Cretaceous period both marsupials and shrew-like insectivores were present. The latter were true placentals and gave rise to most of the orders of mammals we know today.

Towards the end of the era the reptiles became less numerous and in a comparatively short time, from a geological viewpoint, great numbers of species became extinct. These included the dinosaurs, ichthyosaurs and pterosaurs. Nobody knows what caused their extinction, although many theories have been put forward, but the fact remains that by the time the Cenozoic era commenced some 65 million years ago, all but the modern groups of reptiles had disappeared.

The Cenozoic Era (from about 65–2 million years ago)

This era started with great upheavals of the earth's crust forming such mountain ranges as the Alps, Himalayas, Rockies and Andes. During this era our modern flora and fauna developed, with its great diversity of flowering plants, insects, birds and mammals.

This era is described as the age of mammals, because during this period they rapidly diversified to become the dominant vertebrate group. Being warm-blooded they were able to spread to many habitats, and like the reptiles, they showed considerable adaptive radiation, taking over many of the niches which the reptiles had filled in the previous era.

Abundant fossil evidence has been obtained concerning the evolution of mammals during this era, but it is the ancestry of horses that has been worked out in greatest detail (Fig. 15:15).

The earliest type recognisable as a horse, *Hyracotherium*, occurred in the Eocene some 50 million years ago. It was no bigger than a rather small dog but it had relatively long legs. It ran on 3 toes, each encased in a small hoof, but a reduced fourth toe was present on the fore limbs. Its teeth were low-crowned and used for browsing. No doubt it arose from 5-toed ancestors, but fossils of these have either not been found, or have not been recognised as related forms.

Further evolution towards the horses of today showed a gradual increase in size and a

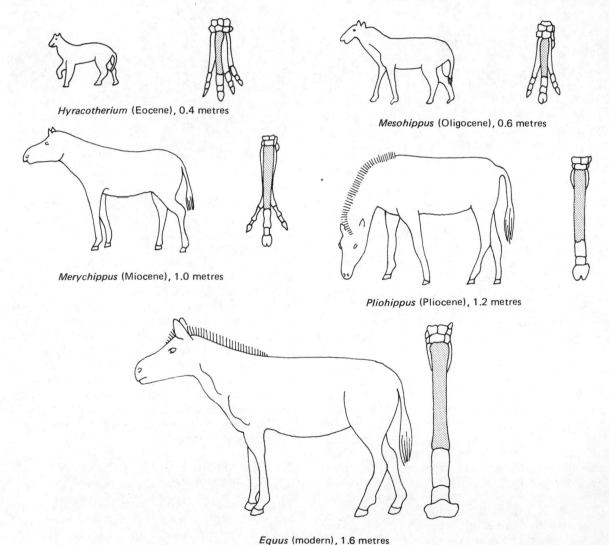

Hyracotherium (Eocene), 0.4 metres

Mesohippus (Oligocene), 0.6 metres

Merychippus (Miocene), 1.0 metres

Pliohippus (Pliocene), 1.2 metres

Equus (modern), 1.6 metres

Fig. 15:15 Diagrams illustrating the evolution of horses (only 5 species shown). The fore feet are not drawn to scale. Figures refer to the shoulder height.

growing tendency for the animal to run on the one central toe, which became much longer and stronger. The other toes became further reduced and in modern types are only tiny vestiges. The teeth also changed, becoming high-crowned and having flat grinding surfaces; this enabled them to graze efficiently.

Horses were very successful, and the fossil record shows that many types were evolved. However, after a time, many of these became extinct. Today we only have three remaining species of the genus *Equus*—the true horses, the asses and the zebras.

At the same time as the horses were evolving, the main mammalian orders were coming into being and by the Pliocene the mammals had reached their zenith. By this time elephant-like mastodons, camels, rhinoceroses, hyaena-like dogs, sabre-toothed tigers, badgers and many of the large herbivores were present. But towards the end of the era, about a million years ago, the climate became much colder and the Ice Age began. Since then the number of mammalian species has steadily declined, and it can now be said that the Age of Mammals has been replaced by the Age of Man.

The evolution of Man

As a result of the most recent fossil finds near Lake Rudolph in East Africa, it appears that man as an upright species with a large brain probably existed 3 million years ago. It is likely, therefore, that his evolution from ape-like ancestors occurred several million years before that.

Man owes his dominant position in the world today primarily to his large brain and his ability to hold objects and use them as tools. His opposable thumb, flexible fingers and sensitivity of touch have played a large part in making this possible. He has also evolved a long life-span and a slow rate of development which have enabled him to learn extensively and gain experience. Also, in his evolution towards a more social life he has developed aspects of parental care and co-operation far greater than in any other species. All these factors have been important in his rise to dominance.

Our most recent evolution has been cultural rather than structural. We not only inherit characteristics which are passed on by genes, but because we have learnt to speak and record we are born into a society with a vast accumulation of knowledge and experience which we can utilise if we have the wisdom to do so.

With the help of science and technology man has built up a considerable degree of dominance over nature; he has conquered many diseases and reached a population density previously unheard of in any species of comparable size which is at the end of a food web. What then of the future?

The future lies with us, but the chief stumbling block towards our further evolution is ourselves. We have still not learnt the art of living together in harmony. Our barbarism is still uncomfortably near the surface of our natures. Our aggression which has served us so well in the past is still with us, and is now a potential danger to our very existence, as its ultimate expression is war. Man has now reached a stage when he can destroy himself—and his future, more than ever before, depends upon whether he can evolve into a new unselfish type of man capable of tolerance, sympathy and generosity between individuals, factions and nations. Perhaps it is because the challenge is so great that it can be exciting to be alive.

Appendix

Books

Broadhurst, P. L. (1963). The Science of Animal Behaviour. Penguin.

Carthy, J. D. (1966). The Study of Behaviour. Edward Arnold.

Cloudsley-Thompson, J. L. and Sankey, J. (1961). Land Invertebrates. Methuen. (A fairly advanced reference book.)

Dale, A. (1951). Patterns of Life. Heinemann.

Devon Trust for Nature Conservation (1972). School Projects in Natural History. Heinemann.

Guilcher, J. M. (1961). The Hidden Life of Flowers. Oliver and Boyd.

Harrison, R. J. (1958). Man the Peculiar Animal. Penguin.

Jackson, R. M. and Raw, F. (1966). Life in the Soil. Edward Arnold.

Lack, D. (1946). Life of the Robin. Witherby. (Also published as a Collins Fontana paperback.)

Lorenz, K. (1952). King Solomon's Ring. Methuen. (A classic study of animal behaviour.)

Mackean, D. G. (1968). An Introduction to Genetics. John Murray.

Mason, A. Stuart (1960). Health and Hormones. Penguin.

Milne, L. J. and Milne, M. The Senses of Animals and Men. Penguin.

Neal, E. G. (1958). Woodland Ecology. Heinemann.

Nuffield Science Teaching Project (1966): Keys to Small Organisms in Soil, Litter and Water Troughs. Longman, Penguin Books.

Rhodes, F. H. T. (1962). The Evolution of Life. Penguin.

Sankey, J. (1958). A Guide to Field Biology. Longman.

Walker, K. (1956, revised). Human Physiology. Penguin.

Ward, B. and Dubos, R. (1972). Only One Earth. André Deutsch. (Also published as a paperback by Penguin.)

8 mm Loop Films

The Heart in Action. Macmillan.

Blood Circulation. Macmillan.

Phagocytosis by a White Blood Corpuscle. Gateway.

Cell Division—Mitosis. Macmillan.

The Eye—Accommodation. Rank.

The Eye—Correction of Short Sight. Rank.

The Eye—Correction of Long Sight. Rank.

The Ear—Structure in Relation to Function. Rank.

The Ear—Perception of Sound Waves. Rank.

Nerve Action—The Reflex Arc. Rank.

The Kidney. Macmillan.

The Human Skin. Macmillan.

Soil Animals. BBC Publications.

Decay of Leaves. BBC Publications.

Drosophila. Gateway.

16 mm Films

Water in Biology. Unilever. (Covers many topics; useful for revision.)

World of the Soil. P.F.B.

The River Must Live. P.F.B.

Shadow of Progress. P.F.B.

Addresses:

BBC Publications, 35 Marylebone High Street, London W1M 4AA.

Gateway Productions Ltd., Waverley Road, Yate, Bristol BS17 5RB.

Macmillan and Co. Ltd., Brunel Road, Basingstoke, Hants.

Petroleum Films Bureau, 4 Brook Street, London W1Y 2AY.

Rank Film Library, 1 Aintree Road, Perivale, Greenford, Middlesex.

Unilever Film Library, Unilever House, P.O. Box 68, London EC4P 4BQ.

Additional notes for teachers

Chapter 5, p. 58. Details for making up pyrogallol are given in the Appendix in Book 1.

p. 64. Molar hydrochloric acid can be prepared by carefully adding 93 cm³ of concentrated acid to 907 cm³ of water.

Chapter 14, p. 169. Plastic "poppet" beads are very suitable for simulating breeding experiments and may be obtained from leading biological suppliers.

p. 174. Ebony-bodied *Drosophila* have been suggested as suitable recessive mutants as they are hardy and breed well. However, some difficulty may be experienced in identifying newly-hatched specimens as it takes a little time for the dark body colour to become apparent. As an alternative vestigial-winged flies could be used, but they are not quite as robust as the ebony-bodied flies.

Acknowledgments

For permission to reproduce photographic illustrations, acknowledgment is due as follows: Fig. 15:5, M. Allman; 6:5, Vivian Almy; 13:11a (I. Beames), 10:10 (I. and L. Beames), 10:1 (Dr G. J. Broekhuysen), 13:11c (S. Gooders), 15:10 (J. Green), 10:8 (Rolf Richter), 11:8 (B. L. Sage), 13:11b (R. T. Smith), 11:6, 15:6 (A. Warren), Ardea Photographics; 4:14, Dr C. G. Butler (from *The World of the Honey Bee*, New Naturalist, Wm. Collins Ltd); 10:14, David Bygott; 13:2 (T. D. Duke), 13:10 (Ken Lambert), Camera Press Ltd; 10:7, 13:4 (Jane Burton), 14:1 (John Markham), 15:9 (Graham Pizzey), 13:11d (Arne Schmitz), 10:5d (D. and K. Urry), Bruce Coleman Ltd; 9:10, John East; 13:12, H. M. Frawley; 13:3 (Bruce Malcolm), Geographical Magazine; 1:1, 1:11, 2:5, 3:5, 3:6, 3:9, 3:10, 4:15, 4:18, 5:6, 5:9, 5:10, 5:11, 6:3, 6:7, 7:1, 7:6, 7:15, 9:5, 14:3, 14:11, Philip Harris Biological Ltd; 13:8, I.C.I. Agricultural Division; 15:2, Dr H. Kettlewell; 10:5a, 10:5e, 10:13, 10:15, 15:13, Geoffrey Kinns; 15:1, Mansell Collection; 2:10, Ministry of Agriculture, Fisheries and Food, Pest Infestation Control Laboratory (Crown Copyright Reserved); 13:5, Andrew Neal; 13:7, A. E. McR. Pearce; 15:12, 15:14, Smithsonian Institution; 7:7, 7:8, 7:9, 9:9, St Mary's Hospital Medical School; 10:5f, Maurice Tibbles; 13:6, John Topham Ltd; 7:10, 9:8, C. James Webb; 14:15, Professor M. H. F. Wilkins; 10:16, Zoological Society of London. Also to the following for other illustrations: 7:16, 7:18, based on Margaret Rutherford, *Man with Two Environments*, Longman Green & Co Ltd; 7:17, based on Clarke et al, *Biology by Inquiry*, Heinemann; 8:3, based on H. G. Q. Rowett, *Basic Anatomy and Physiology*, John Murray Ltd; 10:4, from J. D. Carthy, *The Study of Behaviour*, Edward Arnold Ltd, by kind permission of Professor N. Tinbergen; 11:1, from *Species and Populations*, The Open University Press; 11:2, 15:4c, based on Moon, Otto and Towle, *Modern Biology*, Holt, Rinehart and Winston Inc; 14:12b, from Nuffield Biology Teachers' Guide V, after B. J. F. Haller; 15:15, based on De Beer, *An Atlas of Evolution*, Nelson. We are grateful to André Deutsch Ltd for permission to quote (p. 162) from *Only One Earth* by Barbara Ward and René Dubos.

Index

Figures in bold type show that the subject is illustrated, but in addition there may be a further reference in the text on the same page.